THE NATURE OF PLANETS, DWARF PLANETS, AND SPACE OBJECTS

THE NATURE OF PLANETS, DWARF PLANETS, AND SPACE OBJECTS

Edited by Michael Anderson

Published in 2012 by Britannica Educational Publishing
(a trademark of Encyclopædia Britannica, Inc.)
in association with Rosen Educational Services, LLC
29 East 21st Street, New York, NY 10010.

Distributed exclusively by Rosen Educational Services.
For a listing of additional Britannica Educational Publishing titles, call toll free (800) 237-9932.

First Edition

Britannica Educational Publishing
Michael I. Levy: Executive Editor, Encyclopædia Britannica
J.E. Luebering: Director, Core Reference Group, Encyclopædia Britannica
Adam Augustyn: Assistant Manager, Encyclopædia Britannica

Anthony L. Green: Editor, Compton's by Britannica
Michael Anderson: Senior Editor, Compton's by Britannica
Sherman Hollar: Associate Editor, Compton's by Britannica

Marilyn L. Barton: Senior Coordinator, Production Control
Steven Bosco: Director, Editorial Technologies
Lisa S. Braucher: Senior Producer and Data Editor
Yvette Charboneau: Senior Copy Editor
Kathy Nakamura: Manager, Media Acquisition

Rosen Educational Services
Heather M. Moore Niver: Editor
Nelson Sá: Art Director
Cindy Reiman: Photography Manager
Matthew Cauli: Designer, Cover Design
Introduction by Heather M. Moore Niver

Library of Congress Cataloging-in-Publication Data

The nature of planets, dwarf planets, and space objects / edited by Michael Anderson. -- 1st ed.
p. cm.—(The solar system)
"In association with Britannica Educational Publishing, Rosen Educational Services."
Includes bibliographical references and index.
ISBN 978-1-61530-517-9 (library binding)
1. Planets—Juvenile literature. 2. Satellites—Juvenile literature. 3. Interstellar matter—Juvenile literature. I. Anderson, Michael.
QB602.N38 2012
523.2—dc22

2011001380

Manufactured in the United States of America

Cover, back cover, pp. 3, 27, 32, 33, 54, 60, 68, 84 Shutterstock.com; pp. 10, 29, 39, 58, 66, 79, 88, 89. 91, 94, 95 © www.istockphoto.com/Felix Möckel; interior background © www.istockphoto.com/David Birkbeck

Contents

Introduction 6

Chapter 1 What Is a Planet? 10

Chapter 2 Motions of the Planets 29

Chapter 3 Dwarf Planets 39

Chapter 4 Asteroids 58

Chapter 5 Comets 66

Chapter 6 Meteors and Meteorites 79

Conclusion 88
Glossary 89
For More Information 91
Bibliography 94
Index 95

Introduction

The skies hold many marvels. Among the most fascinating are planets, dwarf planets, and space objects, such as asteroids, comets, and meteors. Astronomers view these celestial bodies through powerful telescopes and send out space probes and satellites to record data and take photographs. These pages will take you on your own exciting journey among planets and space objects.

Planets have long been known to be large, natural bodies that orbit around the Sun, but the specifics of the definition have changed over the years. The most recent definition includes eight planets in the solar system: Mercury, Venus, Earth, and Mars (known as inner planets), and Jupiter, Saturn, Uranus, and Neptune (outer planets). From 1930 until 2006, Pluto was considered a planet, too. But after scientists discovered several other distant bodies similar to Pluto orbiting the Sun, Pluto was reclassified as a dwarf planet.

Most astronomers believe that the solar system started developing around 4.6 billion years ago from a large cloud of gas and dust called the solar nebula. Gravity caused the matter to contract and the cloud began to spin, flattening into a disk. The Sun formed in the center, and the planets formed from remaining material spinning around the Sun. Collisions between the forming planets and smaller bodies may have caused impact craters on their surfaces and even the formation of Earth's Moon.

Dwarf planets are smaller than planets and have not cleared away icy and rocky debris from around their orbits. Cold, dark, distant Pluto, named after the Roman god of the underworld, is probably the best-known dwarf planet. It is part of what is known as the Kuiper belt, a remote region with many small, icy bodies orbiting the Sun. The dwarf planets Eris, Makemake, and Haumea also belong to the Kuiper belt. Ceres, another dwarf planet, orbits the Sun from within the main asteroid belt.

Asteroids are large pieces of rock and metal that orbit the Sun, usually in the zone between the orbits of Mars and Jupiter. Small asteroids regularly strike Earth's surface, but large asteroids crash much less frequently. In the past, such impacts may have caused earthquakes or sea waves—and maybe even the extinction of the dinosaurs.

Comets are another type of space object. The only permanent part of a comet is its

Comet Hale-Bopp. Jamie Cooper/SSPL via Getty Images

nucleus, or core. The nucleus, made up of ice and dust, is commonly called a "dirty snowball." When a comet is near the Sun, it develops a huge cloud of gas and dust, called the coma, as well as one or more dusty or gaseous tails. The comet's head is made up of the nucleus and the coma. The tail can be more than 60 million miles (100 million kilometers) long.

The trails of light called "falling stars" are actually little pieces of metallic or stony material that vaporize upon entering Earth's atmosphere. Before entering they are called meteoroids, and once within Earth's atmosphere they are called meteors. Most meteors vaporize completely before reaching the ground. Sometimes, though, a piece survives and hits the ground. Then it is called a meteorite. Although meteorites consist of materials found on Earth, the proportions are radically different from anything on the planet.

As you watch for meteors or focus a telescope on a distant planet, you see just a tiny fraction of what is in space. Scientists have a battery of technology at their fingertips to study the mysteries of space, but they all started like you: by gazing into the sky with a desire to learn.

1 chapter

What is a Planet?

The relatively large natural bodies that revolve in orbits around the Sun or other stars are called planets. The term does not include small bodies such as comets, meteoroids, and asteroids, many of which are little more than pieces of ice or rock.

The planets that orbit the Sun are part of the solar system, which includes the Sun and all the bodies that circle it. The Sun governs the planets' orbital motions by gravitational attraction and provides the planets with light and heat. Ideas about what makes a planet a planet have changed over time. According to the current definition, there are eight planets in the solar system. In order of increasing mean distance from the Sun, the planets are Mercury, Venus, Earth, Mars, Jupiter, Saturn, Uranus, and Neptune. Pluto was considered the solar system's ninth and outermost planet from 1930 until 2006, when it was reclassified as a dwarf planet.

The concept of what defines a planet changed in another way in the late 20th century when astronomers began discovering distant planets that orbited stars other than the Sun. Previously, the only planets that had been known were the planets of the solar system.

Defining the Term

Mercury, Venus, Mars, Jupiter, and Saturn, all of which can be seen without a telescope, have been known since ancient times. The ancient Greeks used the term "planet," meaning "wanderers," for these five bodies plus the Sun and Earth's

The solar system's four inner planets are much smaller than its four outer planets, and all eight are dwarfed by the Sun they orbit. The sizes of the bodies are shown to scale, but the distances between them are not. The sizes given are the approximate diameters of each body at its equator. **Encyclopædia Britannica, Inc.**

Moon, because the objects appeared to move across the background of the apparently fixed stars. In some languages, the names of these seven bodies are still connected with the days of the week. In time astronomers learned more about how celestial objects move in the sky, and they recognized that the Sun, not Earth, is the center of the solar system. The term "planet" came to be reserved for large bodies that orbit the Sun and not another planet, so the Sun and Moon were no longer considered planets.

With the aid of telescopes, modern astronomers have discovered hundreds of thousands of additional objects orbiting the Sun, including more planets. Although Uranus also is sometimes visible with the naked eye, ancient astronomers were unable to distinguish it from the stars. Uranus was first identified as a planet in 1781. The eighth planet, Neptune, was discovered in 1846.

THE QUESTION OF PLUTO

Pluto was initially considered a planet. Indeed, its discovery in 1930 was the result of a major search for a distant ninth planet. Astronomers found that Pluto was unlike the other planets in many ways, including its

composition of rock and ice and its tilted and very elongated orbit. For decades it seemed unique. Beginning in the late 20th century, however, advances in telescope technology allowed astronomers to discover many more small icy objects with elongated orbits that lie beyond Neptune. Along with Pluto, these numerous objects orbit the Sun in a doughnut-shaped zone called the Kuiper belt. At first it seemed that Pluto was significantly larger than all these objects. But a few Kuiper belt objects were found to be nearly as big as Pluto, and one (which was later named Eris) was slightly larger. This raised the question of whether those bodies should be considered planets if Pluto was.

Characteristics of a Planet

To address this issue and the growing controversy swirling around it, the group that approves the names of astronomical objects for the scientific community issued an official definition in 2006. The organization, known as the International Astronomical Union (IAU), defined a planet as a celestial body that meets three conditions. First, it must orbit the Sun and not be a satellite of another planet. Second, it must be massive

enough for its gravity to have pulled it into a nearly spherical shape. Third, it must have "cleared the neighborhood around its orbit," meaning that it must be massive enough for its gravity to have removed most rocky and icy debris from the vicinity of its orbit.

Pluto is not a planet under this definition. It orbits the Sun and is nearly round, but it has not cleared away many small Kuiper belt objects from the area around its orbit. The IAU created a new category, called dwarf planet, for those nearly spherical celestial bodies that orbit the Sun but that have not cleared the area around their orbits and that are not moons. The first bodies to be designated dwarf planets were Pluto, Eris, and Ceres (which is also the largest asteroid).

The IAU definitions were themselves controversial. Some scientists welcomed them as appropriate recognition that Pluto is one of the larger objects in the Kuiper belt. Other scientists criticized the definitions' lack of precision and scientific rigor and called for their revision.

CLASSIFICATION OF THE PLANETS

The eight planets of the solar system can be divided into two groups—inner planets and

An ultraviolet image taken by the Cassini spacecraft shows the A ring of Saturn. The ringlets shown in turquoise have a greater percentage of water ice and less "dirt" than those shown in red. The reddish area at left is the large gap called the Cassini division; it forms the inner boundary of the A ring. The red band at right is the Encke gap. **NASA/JPL/University of Colorado**

outer planets—according to their basic physical characteristics and positions relative to the Sun. The four inner planets are Mercury, Venus, Earth, and Mars. They are also called the terrestrial, or Earth-like, planets. These

relatively small worlds are composed primarily of rock and metal. With densities ranging from nearly four to five-and-a-half times the density of water, the inner planets have solid surfaces. None of these planets has rings, and only Earth and Mars have moons.

The four outer planets are Jupiter, Saturn, Uranus, and Neptune. They are also called the Jovian, or Jupiter-like, planets. All of them are much bigger than the inner planets, and Jupiter is more massive than the seven other planets combined. Unlike the inner planets, the outer planets have no solid surfaces. Their densities are less than twice the density of water. Jupiter and Saturn are composed mostly of hydrogen and helium in liquid and gaseous form, while Uranus and Neptune consist mainly of melted ices and molten rock, as well as hydrogen and helium. Each of the four has a massive atmosphere or surrounding layers of gases. Each outer planet also has a ring system and many moons.

Formation and Evolution

Although the origin of the solar system is uncertain, most scientists believe that it began to develop about 4.6 billion years ago

Saturn's enormous and stunning rings are well known, but all the outer planets, including Uranus, have rings. A near-infrared image of Uranus's taken from Earth at the Very Large Telescope facility in Chile shows the planet, its rings, and several of its moons. Uranus' rings are nearly impossible to see from Earth in visible light, even with the most powerful telescopes. ESO

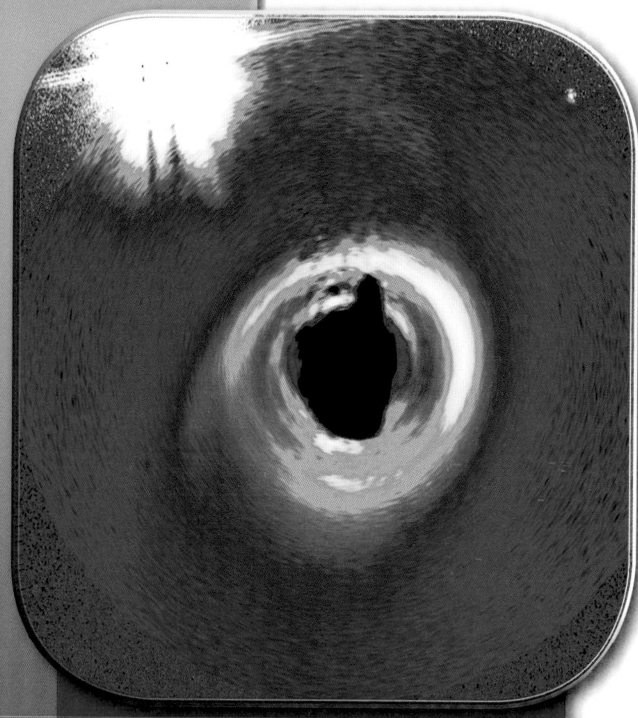

Planets form from disks of gas and dust swirling around new stars. An image taken by the Hubble Space Telescope shows a disk that might be producing planets around a young star named HD 141569A. The star lies some 320 light-years from Earth. The photograph has been modified to simulate what the disk would look like if viewed from above, and false colors were added to better show the disk's structure. To reveal the disk, the star's light was blocked out, so a black central region appears in place of the star. A nearby double-star system appears at upper left. NASA/STScI/ESA

from a large cloud of gas and dust called the solar nebula. The gravity of the cloud began pulling the cloud's matter inward. As the cloud contracted, it began spinning faster and faster and it flattened into a disk. As the material within the cloud compressed, it grew hot. This caused the dust in the cloud to become gaseous.

Most of the cloud's mass was drawn toward the center, eventually forming the Sun. The planets developed from the remaining material—the disk of gas spinning around the forming Sun—as it cooled. This explains why the planets orbit the Sun in nearly the same plane and in the same direction.

Gases in the cooling disk condensed into solid

particles, which began colliding with each other and sticking together. Larger objects began to form. As they traveled around the disk, they swept up smaller material in their paths, a process known as accretion. The larger gravity of the more massive objects also allowed them to attract more matter. Over time, much of the matter clumped together into larger bodies called planetesimals. Ultimately, they formed larger protoplanets, which developed into the planets.

DIFFERENCES BETWEEN INNER AND OUTER PLANETS

The inner and outer planets developed so differently because temperatures were much hotter in the regions near the developing Sun. Close to the center of the solar nebula, the material in the disk condensed into small particles of rock and metal. These particles eventually clumped together into the planetesimals that formed the rocky, dense inner planets.

Farther from the developing Sun, the cooler temperatures allowed not only rock and metal to develop but also gas and the ices of such abundant substances as water, carbon dioxide, and ammonia. The availability

of these ices to the forming outer planets allowed them to become much larger than the inner planets. Eventually, the outer planets grew massive enough for their gravity to attract and retain even the lightest elements, hydrogen and helium. These are the most abundant elements in the universe, so the planets were able to grow enormous. They also developed compositions fairly similar to that of the Sun. Each young outer planet had its own relatively cool nebula from which its regular satellites formed. The irregular satellites are generally thought to be asteroids or other objects that were captured by the planets' strong gravity.

The Role of Collisions

Collisions between the forming planets and large planetesimals and protoplanets probably had dramatic effects. The numerous impact craters on the oldest surfaces of some inner planets and some moons are believed to have been created from such collisions. Astronomers think that Earth's Moon originally may have formed from material scattered by a violent collision of Earth and a protoplanet about the size of Mars. This material may have settled into orbit around

Ancient craters mark the surface of the far side of Earth's Moon, shown in an image taken by the Apollo 16 spacecraft. Many craters are the result of asteroid-sized objects bombarding the Moon during its early history. In fact, most astronomers believe that the Moon formed from fragments created when a larger object slammed into Earth. **F.J. Doyle/National Space Science Data Center**

Earth and accreted to form the Moon. A protoplanet also may have slammed into the developing Mercury and stripped away much of its outer rocky mantle. This would explain why Mercury's core takes up such a large

percentage of the planet's interior. Other protoplanets may have crashed into Venus, greatly slowing its rotation, and Uranus, knocking the planet nearly on its side.

Formation of Layers

As the planets accreted, their interiors grew hot and melted. In a process known

The predominance of water on Earth is apparent in a photograph taken by the Galileo spacecraft in December 1990. The landmass shown is Australia; the body of ice at the bottom is Antarctica. NASA/JPL

as differentiation, heavier materials sank to the centers, generating more heat in the process and, in many planets, gradually forming cores. In the inner planets, the sinking of the heavier materials displaced lighter rocky materials upward, forming mantles of rock. The most buoyant materials rose to the top and solidified into surface crust. Lighter elements escaped from the interiors and formed atmospheres and, on Earth, oceans.

Loss of Internal Heat

Since the planets' formation, many of their physical characteristics have been determined by the manner in which the bodies generated and lost their internal heat. For example, the release of internal heat accounts for the volcanic and tectonic activity that has shaped the crusts of the inner planets. In smaller bodies such as Mercury, Earth's Moon, and many satellites of the outer planets, the internal heat escapes to the surface relatively quickly. As a result, the surface initially undergoes rapid, violent changes. Then, when most of the body's internal heat has dissipated, the surface features stabilize

and remain largely undisturbed as the body ages. Larger bodies such as Earth and Venus lose their heat more slowly. The outer planets are so large that much of their internal heat is still being released.

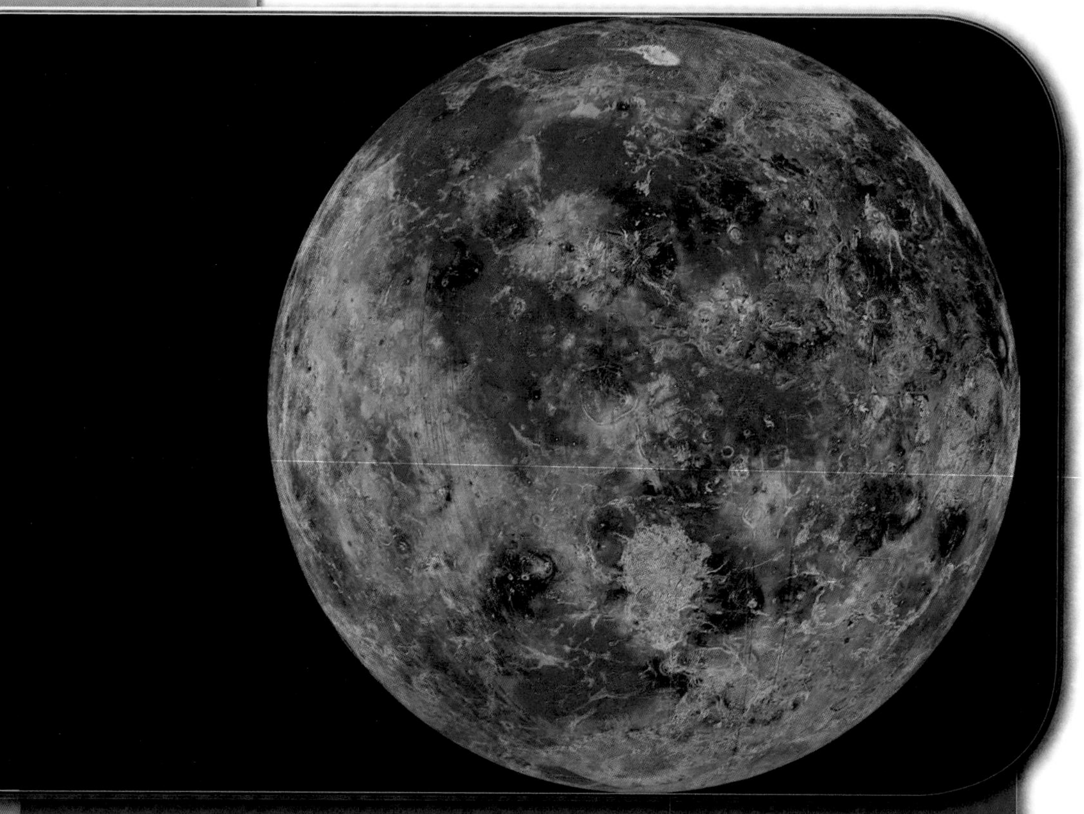

Like the other inner planets of the solar system, Venus has a solid surface. A color-coded radar image reveals its topography. The lowest elevations are colored violet, while the highest elevations appear in red and pink. The large red and pink area at the top is Maxwell Montes, the planet's highest mountain range. The image is centered on 0° longitude, with north at the top, and is based mainly on radar data from the Magellan spacecraft. NASA/JPL/California Institute of Technology

Ongoing Questions

Scientists developed these theories about the formation of the planets based on observations of the solar system. The discovery of planets outside the solar system has challenged some of the details. For example, astronomers have discovered enormous gaseous planets that are closer to their stars than Mercury is to the Sun. This seems to contradict the idea that huge planets can form only in the regions far from the central star. Perhaps these planets initially developed farther away from the star, or perhaps the theories about solar system formation need adjusting in certain ways. The idea that solar systems develop from contracting, spinning clouds of gas and dust, however, is still believed to be correct. Astronomers have observed such disks surrounding several young stars.

Planets Outside the Solar System

People have long wondered whether stars other than the Sun have planets circling them. In the 1990s astronomers found the first evidence that such planets exist. It is now known that there are numerous planets

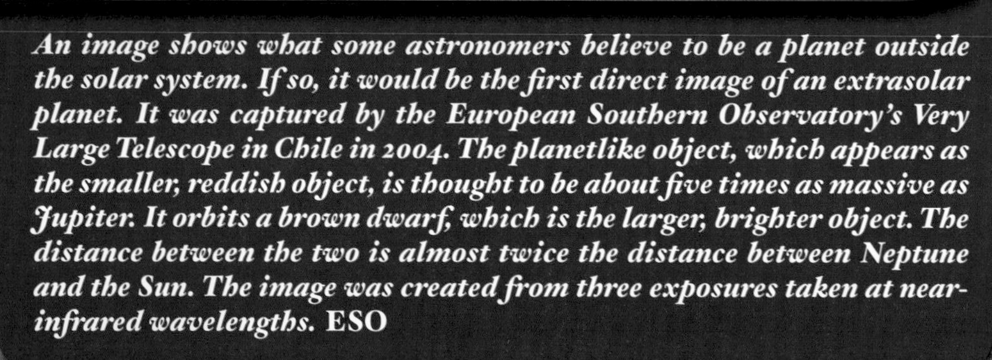

An image shows what some astronomers believe to be a planet outside the solar system. If so, it would be the first direct image of an extrasolar planet. It was captured by the European Southern Observatory's Very Large Telescope in Chile in 2004. The planetlike object, which appears as the smaller, reddish object, is thought to be about five times as massive as Jupiter. It orbits a brown dwarf, which is the larger, brighter object. The distance between the two is almost twice the distance between Neptune and the Sun. The image was created from three exposures taken at near-infrared wavelengths. ESO

in numerous other solar systems. These planets are called extrasolar planets or exoplanets.

Such very distant planets are decidedly difficult to see directly with Earth-based telescopes. They are so far away that they

DISCOVERIES OF EXTRASOLAR PLANETS

The first extrasolar planets were discovered in 1992. They were three planets circling a pulsar—an extremely dense, rapidly spinning neutron star—in the constellation Virgo. In 1995 astronomers announced the first discovery of a planet in orbit around a star more like the Sun, 51 Pegasi. The first confirmed images of an extrasolar planet were released in 2008. Taken with the Hubble Space Telescope, they showed a planet orbiting the star Fomalhaut, only about 25 light-years away from Earth in the constellation Piscis Austrinus. By 2010 more than 400 extrasolar planets had been identified.

would appear extremely small and dark, and they would normally be obscured by the glare of the stars around which they orbit. Also, as seen from Earth, a planet and its parent star are usually too close together for a telescope to optically resolve, or separate, their respective images.

The first planets discovered outside the solar system were identified by several other means. Some methods involve noting the planets' gravitational effects on the observed motion of their parent stars. For instance, an orbiting planet can be detected by the small, periodic wobbles it produces in the parent star's position in space or by deviations in the star's velocity as viewed from Earth. Another method has been used to detect exoplanets that pass directly between their stars and Earth. This causes a kind of eclipse called a transit, allowing astronomers to detect slight dimmings of the star's apparent brightness.

2 chapter

Motions of the Planets

The solar system's planets move through space in two basic ways. They simultaneously orbit the Sun and rotate about their centers.

Orbit

The eight planets orbit the Sun in the same direction as the Sun's rotation, which is counterclockwise as viewed from above Earth's North Pole. The planets also orbit in nearly the same plane, so that their paths trace out a large disk around the Sun's equator. Mercury's orbit is the most tilted. It is inclined about 7 degrees relative to the ecliptic plane, or the plane in which Earth orbits. The orbital planes of all the other planets are within about 3.5 degrees of the ecliptic plane. By comparison, Pluto's orbit is inclined about 17 degrees.

In the early 1600s German astronomer Johannes Kepler discovered three major laws that govern the motions of the planets. The first law describes the shape of their orbits, which are not exactly circular but slightly oval. The orbits form a type of closed curve

Johannes Kepler. **Imagno/Hulton Archive/Getty Images**

called an ellipse. Each ellipse has two imaginary fixed points inside the curve called foci, which is the plural of "focus." (The sum of the distance from a point on the curve to one focus and from the same point to the other focus is the same for all points on the ellipse.) The Sun lies at one of the two foci of the ellipse of each planet's orbit. Mercury has the most eccentric, or elongated, orbit of the planets, while Venus and Neptune have the most circular orbits. The orbits of Pluto, Eris, and many other Kuiper belt objects and comets are significantly more eccentric.

Kepler's second law describes the velocities of the planets in their orbits. It states that an imaginary line drawn from a planet to the Sun sweeps across equal areas in equal periods of time. This means that the planets move faster when their orbits bring them closer to the Sun and more slowly when they are farther away.

Kepler's third law allows one to calculate a planet's orbital period, or the time it takes the planet to complete one orbit around the Sun, if one knows its average distance from the Sun, and vice versa. The law states that the square of a planet's orbital period is proportional to the cube of the planet's average distance from the Sun.

Kepler's Breakthrough

By Johannes Kepler's time, many astronomers had accepted that the Sun was the center of the solar system and that Earth turned on its axis, but they still believed that the planets moved in circular orbits. Because of this, they could not explain the motions of the planets as seen from Earth.

Kepler decided to try explaining these motions by finding another shape for the planetary orbits. Because Mars offered the most typical problem and he had astronomer Tycho Brahe's accurate observations of this planet, Kepler began with it. He tried every possible combination of circular motions in attempts to account for Mars's positions. These all failed, though once a discrepancy of only eight minutes of arc remained unaccounted for. "Out of these eight minutes," he exclaimed, "we will construct a new theory that will explain the motions of all the planets!"

Hampered by poor eyesight and the clumsy mathematical methods of the day, he toiled for six years before finding the answer. Mars follows an elliptical orbit at a speed that varies according to the planet's distance from the Sun. In 1609 he published a book on the results of his work, boldly titling it *The New Astronomy*. He then turned to the other planets and found that their motions corresponded to those of Mars. He also discovered that their periods of revolution—time required to go around the Sun—bore a precise relation to their distances from the Sun. Kepler's great work on planetary motion is summed up in the three principles that have become known as Kepler's laws.

With his new laws established, Kepler could now proceed with his task of revising the *Tabulae Rudolphinae* (Rudolphine Tables), an almanac of the positions of heavenly bodies that, though unsatisfactory, was the best available at the time. Kepler's laws enabled him to predict the positions of the planets by date and hour and have proved to be substantially accurate even to the present day.

ROTATION

In addition to orbiting the Sun, each planet rotates about its axis, an imaginary line that extends through the planet's center and its north and south poles. Most planets rotate in the same direction that they orbit. Only Venus and Uranus rotate in the opposite direction, which is called retrograde rotation. The rotational axes of all the planets except Uranus are more or less upright, or perpendicular, to the ecliptic plane. Oddly, Uranus's axis is almost parallel to the ecliptic plane, so the planet rotates nearly on its side.

YEARS, DAYS, AND SEASONS

A planet's motions in space define the length of a year and of a day on that planet. A year is the time it takes the planet to complete one orbit around the Sun. Earth's orbital period, for example, is about 365 days, so that is the length of one Earth year.

The length of a planet's day is defined by its rotation in two different ways. A day is the time it takes the planet to rotate on its axis once relative to the Sun. Astronomers call this a solar day. During one solar day, the Sun moves from its noontime position in the sky, sets, and rises to the noon position again. A sidereal day is the time it takes the planet to rotate once relative to distant stars. In most cases, the solar and sidereal days are nearly the same length. Earth's solar day is about four minutes longer than its sidereal day, while on Mars and all the outer planets this difference is less than a minute. Mercury and Venus are unusual in that their solar and sidereal days differ by many hours.

Seasons are caused mainly by the tilt of a planet's spin axis relative to its plane of orbit. The variation in the planet's distance from the Sun over the course of a year is a secondary factor. The axes of Mercury, Venus, and

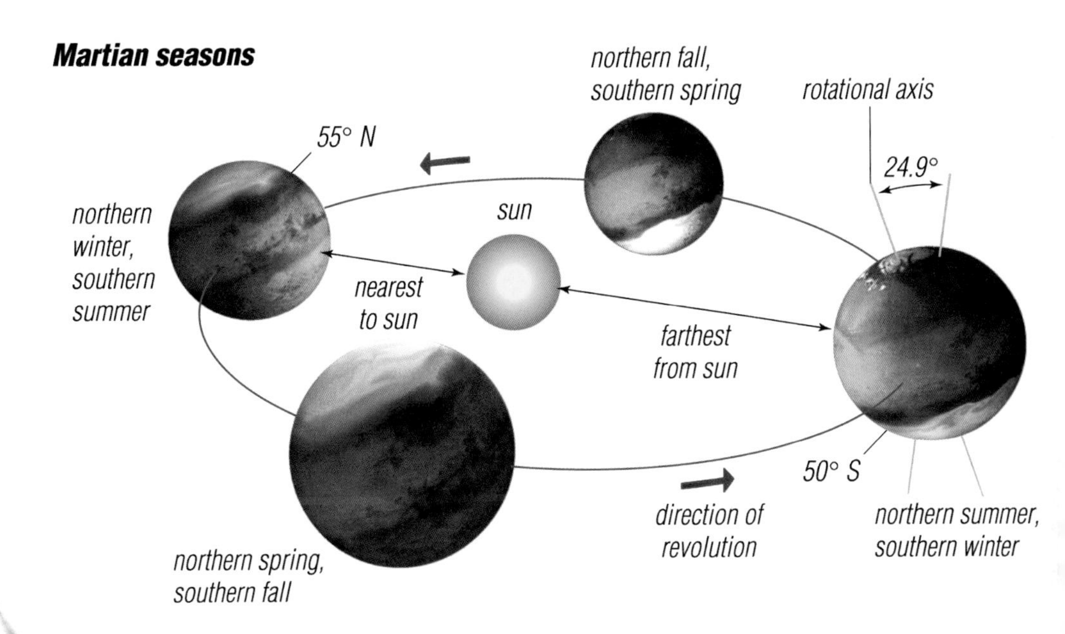

Mars's spin axis is tilted about 24.9° relative to the plane in which it orbits. As the planet travels in its orbit, first the northern hemisphere, then the southern hemisphere is tipped toward the Sun. As a result, there are four distinct seasons on Mars. The ice caps at the poles alternately grow and shrink as the seasons change. **Encyclopædia Britannica, Inc.**

Jupiter are barely tilted, so they experience little or no seasonal differences in solar illumination and weather.

Earth's axis is tilted about 23.5 degrees. As the planet travels around the Sun, first one hemisphere then the other is tipped slightly toward the Sun. As a result, the angle of the

sunlight reaching the ground at noon varies over the course of the year. The half of Earth that is tipped toward the Sun experiences summer, while the half that is tipped away experiences winter. During summer the Sun is higher in the sky at noon and the hours of daylight per day are longer, which causes greater heating.

Mars, Saturn, and Neptune have axial tilts that are a bit greater than Earth's—roughly 25 to 30 degrees—so they also have seasons. Because these planets take longer to orbit the Sun, each season is much longer than on Earth.

The extreme tilt of Uranus's axis, more than 97 degrees, gives it extreme seasons. For part of the year, the planet's north pole points roughly toward the Sun, while the southern hemisphere remains dark day and night. Later in the year conditions are reversed, with the northern hemisphere draped in darkness and the southern hemisphere bathed in light.

APPARENT MOTIONS

The motions of the planets as observed from Earth, called their apparent motions, are complicated by Earth's own revolution, rotation, and slightly tilted axis. Earth

rotates from west to east, so that both the stars and the planets appear to rise in the east each morning and set in the west each night. Observations made at the same time every night will show that a planet usually appears in the sky slightly to the east of its position the previous night. Periodically, a planet will appear to change direction for several nights or more and move slightly to the west of its previous position. However, the planet does not actually change direction along its orbital path. Such an apparent reversal in direction occurs whenever Earth "overtakes" an outer planet. For example, because Saturn requires almost 30 Earth years to complete one orbit around the Sun, Earth often passes between Saturn and the Sun. As this occurs, Saturn appears to gradually slow down and then temporarily reverse its course against the background of stars.

Mercury and Venus also have unique apparent motions. Because their orbits are smaller in diameter than Earth's, both appear to move from one side of the Sun to the other. When either planet appears east of the Sun to observers on Earth, it appears as an "evening star" and when west of the Sun, as a "morning star."

MOTIONS OF THE PLANETS' MOONS

The natural satellites, or moons, revolving around the planets follow the same laws of orbital motion as the planets. Most of the solar system's larger moons and many small moons orbit in the same direction as the planet they circle and roughly in the planet's equatorial plane (the plane that is perpendicular to the planet's rotational axis and passes through the center of the planet). These moons are called regular satellites. In addition, most large moons, including Earth's, rotate on their axes once for each revolution around the planet. As a result, these satellites always show the same side to the planet.

3 chapter

Dwarf Planets

The objects called dwarf planets are similar to the solar system's eight planets but are smaller. Like planets, they are large, roundish objects that orbit the Sun but that are not moons. The category of dwarf planet was created as a result of intense debate as to whether Pluto, which had been considered the solar system's ninth planet, should really be called a planet. In 2006 the International Astronomical Union (IAU) redefined "planet" in such a way that excluded Pluto and, at the same time, established a new class of objects called dwarf planets. The first three objects classified as dwarf planets were Pluto, Eris, and Ceres. Makemake (pronounced "mah-kay mah-kay") and Haumea were named dwarf planets in 2008.

Pluto, Eris, Makemake, and Haumea are large members of the Kuiper belt, a distant region containing countless small, icy bodies orbiting the Sun. Ceres, the largest asteroid, orbits the Sun from within the main asteroid belt. These five bodies do not qualify as planets under the IAU's definition because they do not meet one of the conditions—they

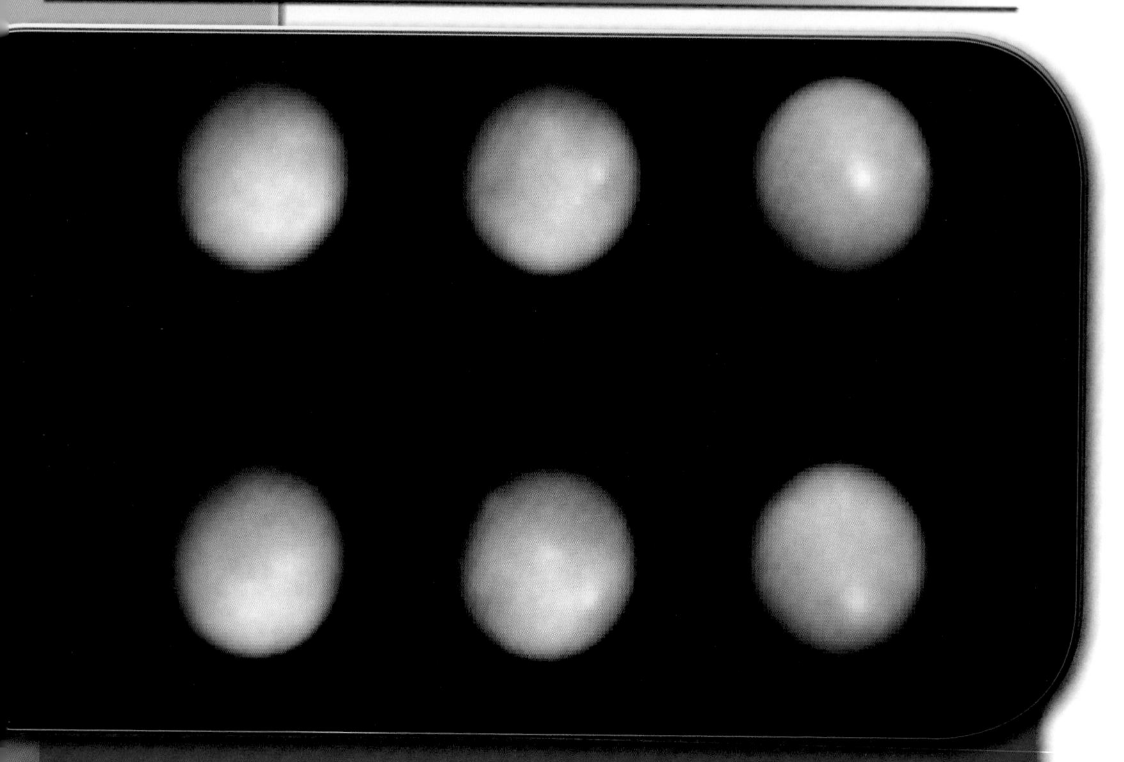

Six false-color images show the rotation of Ceres, a dwarf planet and the largest asteroid. Like the eight planets, Ceres is massive enough for its gravity to have molded it into a round shape. Each image is a composite of many exposures taken by the Hubble Space Telescope in visible and ultraviolet light. Ceres' small size and distance from Earth make it difficult to photograph from Earth's vicinity. **ESA/STScI/NASA**

have not cleared away many chunks of icy and rocky debris from their orbital vicinities. (For more information, see "Defining the Term" in Chapter 1.)

For practical purposes, objects classified as dwarf planets are smaller than the planet Mercury, which has a diameter of about 3,032 miles (4,879 kilometers). Eris is thought to have a diameter of roughly 1,550 miles (2,500 kilometers). Pluto is slightly smaller, with a diameter of about 1,456 miles (2,344 kilometers). With a diameter of perhaps about 1,000 miles (1,600 kilometers), Makemake is thought to be some two thirds the size of Pluto. Ceres is the smallest dwarf planet, with a diameter of about 584 miles (940 kilometers).

Haumea is an unusual object. Although it is substantially rounded, it is also quite elongated. It rotates about its axis so quickly—completing one rotation in just under four hours—that it is pulled into a shape somewhat like an American football. Its longest dimension is thought to be roughly the size of the diameter of Makemake or Pluto.

In 2008 the IAU decided on a name for a new subcategory of dwarf planets, called plutoids. A plutoid is a dwarf planet whose orbit takes it farther from the Sun than Neptune, on average. In addition, to be named as a plutoid, a dwarf planet must meet a requirement for minimum brightness (an absolute magnitude greater than +1). Pluto, Eris, Makemake, and Haumea are considered to be both dwarf planets and plutoids. Ceres, which orbits much closer to the Sun, is a dwarf planet but not a plutoid.

In this chapter, Pluto and Eris will be discussed as representative examples of dwarf planets.

PLUTO: THE PROTOTYPE DWARF PLANET

The best-known dwarf planet is Pluto. For 76 years, however, from its discovery in 1930 until 2006, this distant rocky and icy body was considered the ninth and outermost planet. Pluto is on average about 39.5 times farther from the Sun than is Earth. As a result, very little sunlight reaches Pluto, so it must be a dark, frigid world. Fittingly, it was named after the god of the underworld in ancient Roman mythology. It has three moons, two of which are tiny. Its largest moon, Charon, is so large with respect to Pluto that the two are often considered a double-body system—that is, two objects of similar size revolving around each other.

Pluto is too far from Earth to be visible with the unaided eye. It was discovered in the early 20th century when astronomers began searching the skies for a new planet beyond the eight planets then known. They had noted what seemed to be irregularities in the orbits of the planets Uranus and

Neptune, and they thought that the gravity of a planet beyond Neptune might be the cause. In 1929 the Lowell Observatory hired Clyde Tombaugh, a 23-year-old amateur astronomer, to continue the search. He discovered Pluto in 1930, and it was considered a planet. Pluto turned out to be much too small, however, to account for the orbital disturbances of Uranus and Neptune (which more accurate data later showed not to exist anyway). The discovery of Pluto was thus serendipitous—the result of a lucky accident.

Pluto proved to be unlike any other planet in orbit, size, density, and composition. In the late 20th century, astronomers began to discover other icy objects like

Cold, dark Pluto was named after the Roman god of the underworld.
Archive Photos/Getty Images

Pluto that orbit the Sun in the Kuiper belt, beginning the process that led to Pluto's reclassification as a dwarf planet.

Scientists know less about Pluto than they do about the planets. Because Pluto is so far from Earth, its features are difficult to observe with even the most powerful telescopes. Key observations have been made using instruments that orbit Earth from

An artist's conception shows the New Horizons spacecraft approaching Pluto and its three moons. **NASA/Johns Hopkins University Applied Physics Laboratory/Southwest Research Institute**

above its atmosphere, including the Hubble Space Telescope. Earth-based telescopes outfitted with equipment to reduce the blurring effects of the atmosphere have also been effective. The first spacecraft mission to Pluto, New Horizons, is expected to dramatically increase knowledge about the distant world. The U.S. National Aeronautics and Space Administration (NASA) launched the unmanned probe in 2006 on a nine-year trip to Pluto and Charon.

Basic Astronomical Data about Pluto

Pluto is one of the largest members of the Kuiper belt, but it is considerably smaller than the solar system's planets. In size, density, and composition it most resembles Triton, the large icy moon of Neptune. The diameter at Pluto's equator is about 1,456 miles (2,344 kilometers), which is less than half that of the smallest planet, Mercury. Several moons, including Earth's Moon and Triton, are larger. Triton has about twice the mass of Pluto, and Mercury has more than 27 times the mass. Like Triton, Pluto is only about twice as dense as water. Its low density suggests that it is made of a high

percentage of ice as well as rock. By comparison, Mercury and Earth are both more than five times as dense as water.

ORBIT

Like the eight planets, Pluto revolves around the Sun in an elliptical, or oval-shaped, orbit. However, Pluto's orbit is more tilted and more eccentric, or elongated, than the orbits of the planets. While the planets orbit in about the same plane as Earth does, Pluto's orbit is inclined about 17 degrees from that plane. Mercury has the most tilted orbit of the planets, with an inclination of about 7 degrees.

Pluto's orbit is quite elongated. As a result, as it travels along its path, its distance from the Sun varies considerably, from roughly 2.8 billion to 4.6 billion miles (4.5 billion to 7.3 billion kilometers). For a small part of its orbit, Pluto is actually closer to the Sun than Neptune is. The last time this happened was in 1979–99. Pluto and Neptune will never collide, however. In the time it takes Neptune to revolve around the Sun three times, Pluto revolves only twice, in such a way that the bodies never pass each

other closely. Since Pluto is so distant, it takes nearly 248 Earth years to complete just one trip around the Sun. In other words, a year on Pluto is about 248 times longer than one on Earth.

Rotation

Scientists studying Pluto through telescopes have observed that its brightness level varies regularly in a period of about 6.4 days. This variation indicates that some areas of the surface reflect more light than others and that Pluto completes one rotation on its axis in about 6.4 days. In other words, one day on Pluto lasts about 6.4 Earth days.

Pluto's rotational axis is tipped about 122 degrees relative to its orbital plane, so that, like Uranus, it lies nearly on its side. Both those bodies spin in retrograde motion, or the direction opposite that of most of the planets. An observer on Pluto would see the Sun rise in the west and set in the east.

Seasons

Pluto's large, eccentric orbit and the great tilt of its axis must give it long, uneven, and

extreme seasons. The amount of sunlight reaching the dwarf planet varies greatly during its year, since its distance from the Sun varies so much. Because its axis is so tilted, during parts of its orbit it points its north pole almost directly at the Sun. It is then dark in the southern hemisphere day and night. When Pluto is on the opposite side of its orbit, the situation is reversed, with the northern hemisphere in constant darkness.

Pluto's Atmosphere, Surface, and Interior

In 1988 Pluto passed directly in front of a star as seen from Earth, an event called an occultation. The way the star's light gradually dimmed before being temporarily blocked out completely to Earth-based observers showed scientists that Pluto has an atmosphere, or surrounding layer of gases. It is quite tenuous and extends far above the dwarf planet.

The atmosphere is composed of nitrogen, with smaller amounts of methane and carbon monoxide. The density of the gases is always low, but it must vary as Pluto's

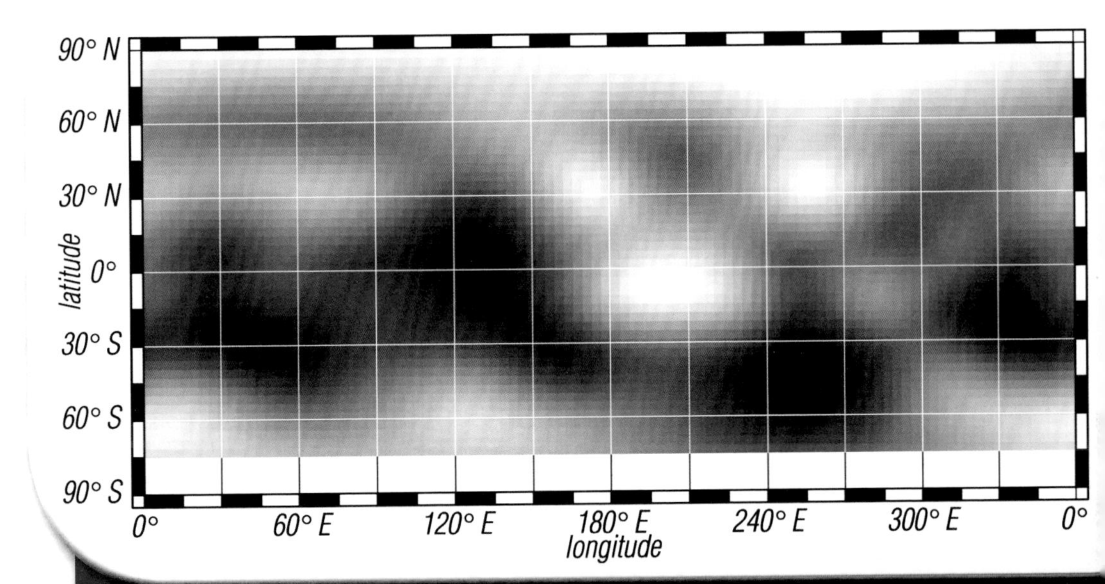

Bright and dark regions on Pluto's surface appear in a map based on images taken by the Hubble Space Telescope. The north polar region generally has bright areas, while the equatorial region, particularly to the south, has more dark patches. The map is a Mercator projection; the blue tint was added in reproduction. **Encyclopædia Britannica, Inc.**

distance from the Sun varies. At its closest to the Sun, Pluto is warmer. Some frozen gases on its surface must then vaporize, becoming gases in the air. As Pluto moves away from the Sun and gets colder, gases in the air must freeze onto the surface again. Scientists estimated that in 2000 the surface pressure was only a few microbars to

several tens of microbars, with one microbar being about a millionth of the surface pressure on Earth. Pluto was then near its closest to the Sun, so the atmosphere was near its densest. The atmosphere may not be detectable at all when Pluto is farthest from the Sun.

Pluto's surface has both bright and dark regions. Overall, it reflects about 55 percent of the light that reaches it. In comparison, Earth's Moon reflects only about 10 percent of the light it receives, while icy Triton reflects about 80 percent, as ice is highly reflective. Pluto's fairly high reflectivity suggests that its surface consists partly of ices and partly of something else. The brighter regions seem to be mostly frozen nitrogen, with some frozen methane, water ice, and frozen carbon monoxide. The area around Pluto's south pole is especially bright. Little is known about the darker regions of the surface, which are somewhat reddish. It is thought that they contain some mixture of organic compounds.

Because it is so far from the Sun, Pluto receives only about $\frac{1}{1,600}$ of the amount of sunlight that Earth does, on average. Its surface is extremely cold. Different types of observations have suggested that the

surface temperature may be about -355 °F to -397 °F (-215 °C to -238 °C). The temperature probably varies seasonally, and the brighter areas must be generally colder than the darker ones.

Pluto is thought to be made of more than half rock, with the rest ice, probably water ice. Scientists think its interior may have separated into layers, with a rocky core surrounded by a mantle of water ice, but more information is needed.

The Moons of Pluto

In the years after Pluto's discovery, attempts to detect moons were unsuccessful because Pluto is so remote. Moreover, Charon, its only major moon, lies unusually close to Pluto. It is only about 12,200 miles (19,640 kilometers) from Pluto's center, which is nearly 20 times closer than Earth's Moon is to Earth. As a result, Charon was obscured by the glare of Pluto's light. In 1978 James W. Christy and Robert S. Harrington of the U.S. Naval Observatory discovered Charon while examining photographs of Pluto. The moon was named after the boatman in ancient Greek mythology who ferried souls across

Pluto and its three known moons, Charon, Nix, and Hydra, appear in an image taken by the Hubble Space Telescope. HST Pluto Companion Search/ESA/NASA

the river Styx to Hades (the Greek counterpart of Pluto) for judgment. (Codiscoverer Christy also had his wife, Charlene—nicknamed "Char"—in mind in naming the moon and wanted its initial sound to be pronounced "sh" rather than "k.")

For a moon, Charon is rather large relative to Pluto. Its diameter is about 750 miles (1,210 kilometers), which is about half that of Pluto. Its density is approximately 1.7 times that of water, and it reflects about 35 percent of the light that hits it. It is thought to be a bit more than half rock, with the rest ice. Unlike Pluto, Charon probably has water ice covering much of its surface. Scientists were able to refine their estimates of Charon's size and density after they observed it occulting a star in 2005. The occultation also indicated that the moon either does not have an atmosphere or has an extremely tenuous one (much less massive even than Pluto's).

Charon completes one revolution around Pluto and one rotation on its axis in about 6.4 days. It thus rotates synchronously, so that the same hemisphere is always facing Pluto. Because Pluto's rotational period also is about 6.4 days, Pluto in turn always faces its same hemisphere toward Charon.

Formation of Pluto's Moons

Scientists think that Pluto formed when the solar system condensed out of a gaseous cloud some 4.6 billion years ago. A collision between Pluto and a large body is thought to have knocked debris into a ring around Pluto, and the debris clumped together into an object that ultimately developed into Charon. Earth and its large Moon probably formed through a similar collision.

A team of astronomers discovered two much smaller moons in 2005 in images taken with the Hubble Space Telescope. The moons, named Nix and Hydra, may be roughly 35 miles (55 kilometers) in diameter. They may have formed from the same collision that created Charon.

Beyond Pluto: The Dwarf Planet Eris

The object named Eris orbits the Sun from well beyond the orbits of Neptune and Pluto. It is the largest-known member of the Kuiper belt and is classified as a dwarf planet. In addition, as a large, bright Kuiper belt object, it is also a plutoid—a bright dwarf planet whose orbit takes it farther from the Sun than Neptune.

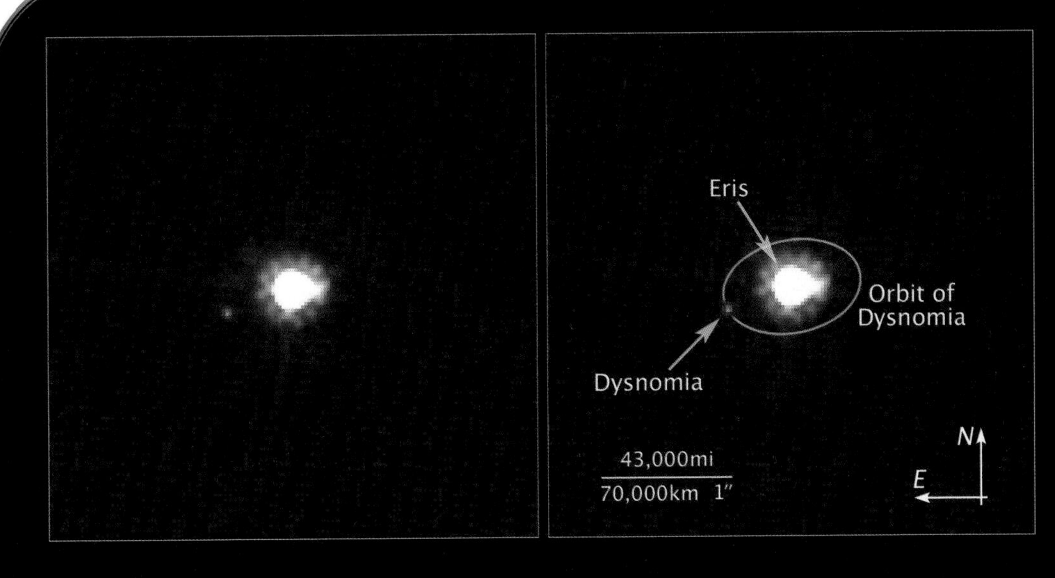

The dwarf planet Eris is the largest-known object in the Kuiper belt. Its one known moon is named Dysnomia. **NASA, ESA, and M. Brown (California Institute of Technology)**

BASIC ASTRONOMICAL DATA

Eris's diameter of roughly 1,550 miles (2,500 kilometers) makes it slightly larger than Pluto. Eris is even more remote than Pluto, being on average nearly 68 times farther from the Sun than is Earth, compared with 39.5 times for Pluto. Scientists calculate that

it takes Eris some 560 Earth years to complete just one revolution around the Sun. Its orbit is highly eccentric, or elongated, and extremely tilted. Its orbit is so tilted that some scientists call it a member of the Kuiper belt's scattered disk rather than of the Kuiper belt itself. The surface of Eris may be covered with white methane ice. Eris has one known moon, Dysnomia. The moon is about an eighth as big as Eris and takes about two weeks to circle the dwarf planet.

DISCOVERY

Eris was discovered in 2005 in images taken two years earlier at Palomar Observatory in the U.S. state of California. Its discoverers were astronomers Michael E. Brown, Chad Trujillo, and David Rabinowitz. The Kuiper belt object was provisionally designated 2003 UB313 and was nicknamed "Xena" (after a character in a television series) and "the 10th planet" before receiving its official name, Eris, in 2006. Eris was named after the goddess of discord and strife in ancient Greek mythology, who was best known for starting the Trojan War. The name is fitting, because the discovery of the celestial object led to a great controversy in the world of planetary science. Because Eris is larger than Pluto, some scientists thought that it, too, should be considered one of the solar system's major planets. However, the new classification scheme adopted in 2006 excluded both Pluto and Eris from the planet category.

Eris's moon, Dysnomia, was discovered in 2005 in infrared images taken at the W.M. Keck Observatory in the U.S. state of Hawaii. The moon was named for the daughter of Eris in Greek mythology, who was associated with lawlessness.

4 chapter

Asteroids

The many small bodies called asteroids are chunks of rock and metal that orbit the Sun. Most are found in the main asteroid belt, a doughnut-shaped zone between the orbits of Mars and Jupiter. Jupiter is the largest planet by far. Astronomers think that when the solar system was forming, the immense pull of gravity from the object that became Jupiter prevented the asteroids from clumping together to form a planet.

The discovery of asteroids dates to 1801, when the Italian astronomer Giuseppi Piazzi observed an object that he later named Ceres. It is the largest known asteroid. Astronomers identified hundreds more asteroids in the 1800s and tens of thousands more in the 1900s. The majority of known asteroids have been discovered since the late 1990s.

Physical Characteristics

Along with meteoroids and comets, asteroids belong to a large group of space objects called small bodies. This label is given to any natural solar system object other than the Sun, a

Numerous craters mark the surface of Gaspra, an asteroid of the main belt, which appears in a composite of two images taken by the Galileo spacecraft. The colors were enhanced by computer to highlight subtle variations in surface properties. **NASA/JPL/Caltech**

planet, a dwarf planet, or a moon. Asteroids are also called minor planets because they are smaller and much more numerous than the major planets of the solar system. Ceres is the largest asteroid by far, with a diameter of about 584 miles (940 kilometers). It is massive enough to also be considered a dwarf planet. Next in size are Vesta and Pallas, at roughly 330 to 310 miles (530 to 500 kilometers). Only about 30 asteroids are greater than 125 miles (200 kilometers) in diameter. Most

TROJANS

Most asteroids occur in groups within the main belt. However, some asteroids, called Trojans, orbit in two clusters at Jupiter's distance from the Sun, with one group at about 60 degrees ahead of Jupiter and the other about 60 degrees behind. About 40 Trojans are known. They are quite dark, like C-type asteroids, but many have a distinctly dark-reddish hue. They are rich in organic compounds, may have ice deep within, and are thought to resemble comet nuclei.

The Trojan asteroids are named for heroes of Greece and Troy in *The Iliad*, a work by the ancient Greek poet Homer. The first was Achilles, discovered by Max Wolf in 1906. Of the named Trojans, Achilles, Hector, Nestor, Agamemnon, Odysseus, Ajax, Antilochus, Diomedes, and Menelaus are in the group ahead of Jupiter. Patroclus, Priamus, Aeneas, Anchises, and Troilus are in the group behind Jupiter.

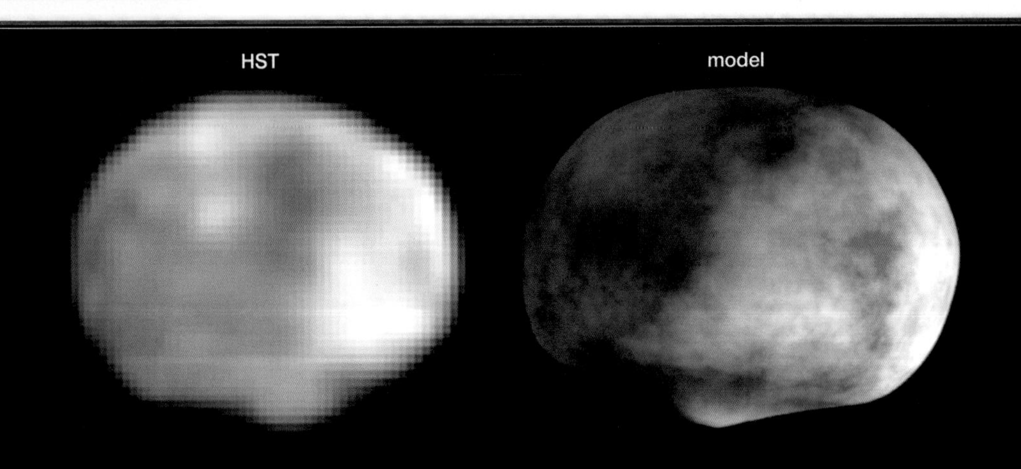

are much smaller. There are probably millions of boulder-sized asteroids in the solar system. These small objects likely result from collisions of larger asteroids.

The largest asteroids have enough mass for their gravity to have pulled them into nearly spherical shapes. Smaller asteroids have a wide range of shapes, including elongated and very irregular forms. Eros, for example, is shaped somewhat like a potato,

Vesta, one of the largest asteroids, has a mostly rounded shape. Earlier in its history, however, a collision with another body gouged out a huge basin near its south pole, shown at bottom in the images. The asteroid appears in three views based on data from the Hubble Space Telescope: an image (left) *taken by the telescope, a computer-generated model* (center)*, and a color-coded image* (below) *that shows differences in elevation, with pink showing the highest elevations and blue showing the lowest. The elevation map is tilted somewhat relative to the two other views.* **Source: Ben Zellner, Georgia Southern University; Peter Thomas, Cornell University; NASA © Encyclopædia Britannica, Inc.**

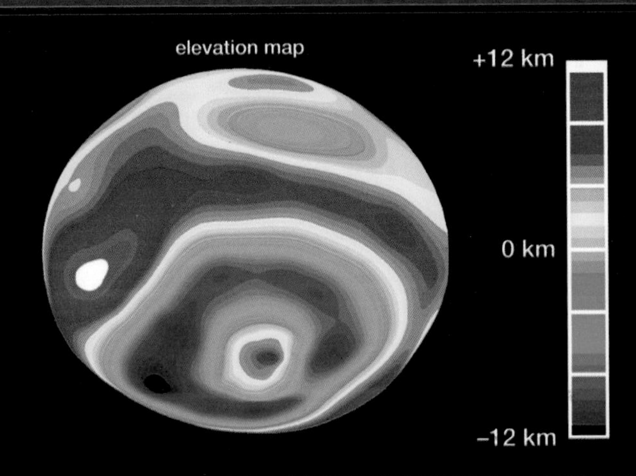

with dimensions of approximately 8 × 8 × 20.5 miles (13 × 13 × 33 kilometers).

Astronomers classify asteroids into more than a dozen categories by the fraction of starlight they reflect and by the spectrum (or color) of the reflected light, which provide clues to the surface composition. There are three broad types: C, S, or M. Most large- and medium-sized asteroids are C-type, or carbonaceous, asteroids. They are very dark, containing a considerable amount of carbon and complex organic compounds. C-type asteroids predominate in the outer part of the main belt. S-type, or silicaceous, asteroids are relatively bright stony bodies. Their surfaces contain silicate rock with some iron. S-type asteroids are found mainly in the inner regions of the main belt. M-type, or metallic, asteroids also are fairly bright. Their surfaces probably contain much iron. They occur in the middle of the main belt.

Near-Earth and Potentially Hazardous Asteroids

Other asteroids approach the Sun more closely than do those in the main belt. Those that pass close to Earth's orbit—within roughly 28 million miles (45 million kilometers)—are called near-Earth asteroids. The orbits of some of these asteroids cross Earth's orbit, so one could collide with Earth in the future. Several programs are dedicated to identifying and studying these potentially hazardous asteroids.

Small asteroids and asteroid fragments regularly strike Earth's surface in the form of meteorites. Much less often, large asteroids crash into Earth, forming huge craters. Past large impacts may have caused earthquakes, giant sea waves, and even global dust clouds that blocked sunlight for long periods. One theory put forth to explain the mass extinction of the dinosaurs and other species on Earth some 65 million years ago is that a large asteroid slammed into Earth, causing a massive disturbance in the global climate. A crater at Chicxulub, in southeastern Mexico, is thought to have been created by that asteroid.

Spacecraft Exploration

The first spacecraft to encounter and photograph an asteroid up close was NASA's unmanned space probe Galileo. The craft flew by two S-type main-belt asteroids—Gaspra in 1991 and Ida in 1993. Ida was found to have a moonlet, Dactyl, which measures about a mile

(1.5 kilometers) across. Both Ida and Dactyl appear to have the composition of ordinary chondrite meteorites, the most common kind in meteorite collections. Dactyl was the first natural satellite (moon) to be discovered orbiting an asteroid. Since its discovery, about 20 percent of near-Earth asteroids have been found to have moons or to be double-body systems. Smaller percentages of main-belt and Trojan asteroids have moons.

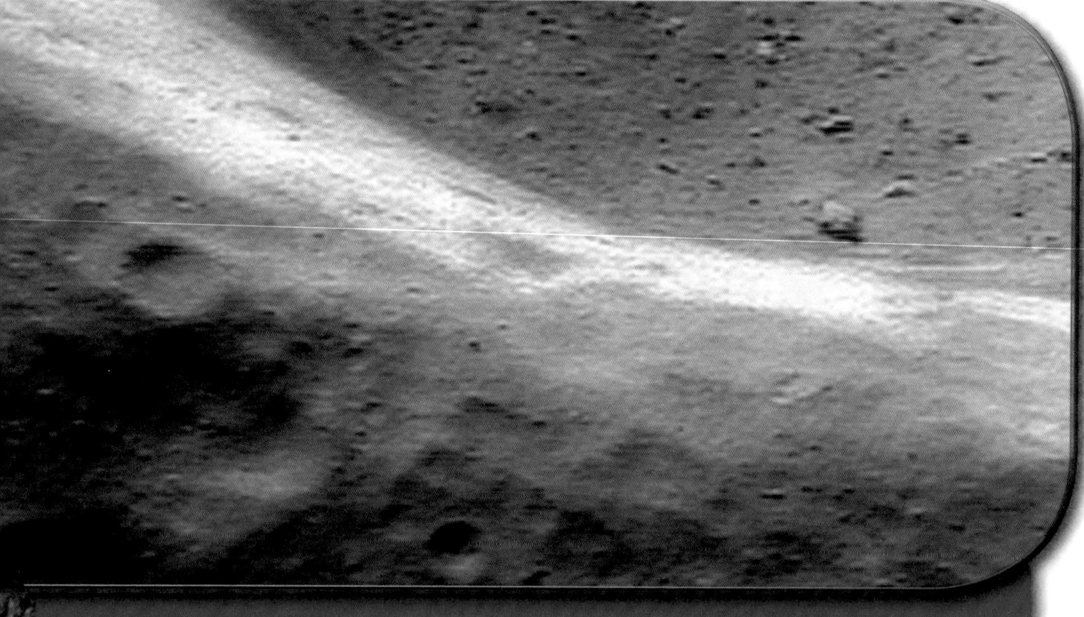

A false-color close-up of the asteroid Eros shows dust and fragments of rocky debris inside a large crater. The NEAR Shoemaker spacecraft took the image from about 30 miles (50 kilometers) above the asteroid's surface. The redder areas have been chemically altered by small impacts and exposure to the solar wind. The bluer areas have been less "weathered."
NASA/The Johns Hopkins Unviersity Applied Physics Laboratory

NASA's Near Earth Asteroid Rendezvous (NEAR) Shoemaker probe was the first craft to fly past a C-type asteroid, Mathilde, in 1997. It later became the first craft to conduct a long-term study of an asteroid at close range, orbiting the S-type near-Earth asteroid Eros for about a year in 2000–01. The first craft to land on a small body, it made a controlled descent onto Eros's surface in 2001. One important discovery was that Eros never underwent extensive melting and separation into layers, so it may be a pristine sample of primordial solar system material.

Other spacecraft to fly past asteroids include NASA's Deep Space 1, which passed asteroid Braille in 1999, and NASA's Stardust, which flew by asteroid Annefrank in 2002. In 2005 the Japanese craft Hayabusa mapped the surface of asteroid Itokawa before briefly touching down on the asteroid a couple of times to try to collect a sample of surface material. However, control and communications problems delayed the craft's return to Earth. In 2007 NASA launched its Dawn mission to Vesta and Ceres. The European Space Agency's Rosetta probe flew by the asteroids Šteins in 2008 and Lutetia in 2010 on the way to its main mission, to orbit Comet 67P/Churyumov-Gerasimenko.

5 chapter

Comets

When near the Sun, the small bodies called comets develop a hazy cloud of gases and dust. They also often develop long, glowing tails. However, a comet exists as only a small core of ice and dust for most or even its entire orbit around the Sun. Comets can be easily seen from Earth only when they approach the Sun closely. Even then, most are visible only with a telescope. Among the exceptionally bright "naked eye" comets seen from Earth after 1900 were the Great Comet of 1910, Halley's, Skjellerup-Maristany, Seki-Lines, Ikeya-Seki, Arend-Roland, Bennett, West, Hyakutake, Hale-Bopp, McNaught, and Holmes. When comets are far from the Sun, they appear in large telescopes as a point of light, like a star.

Most comets originate in the extremely distant, outer regions of the solar system. They are thought to be the remnants of the building blocks that produced the planets Uranus and Neptune some 4.6 billion years ago. Comets remain essentially unchanged

Comet Lulin. **ChinaFotoPress/Getty Images**

when they are away from the Sun in the deep cold of space, which for many comets can be for eons. For this reason, astronomers think that comets may contain some of the oldest and best-preserved material in the solar system.

Halley's Comet

Many people look forward with interest to sighting a comet, but for many centuries comets were believed to have an evil influence on human affairs. In particular, they were thought to foretell plagues, wars, and death. It was not until the 17th century that they began to be properly understood. Astronomer Edmond Halley studied the written accounts of 24 comets that had been seen from 1337 to 1698 and calculated their orbits. He found that the comets of 1531, 1607, and 1682 moved in almost the same paths, and he concluded that they were all the same comet, which would return in about 1758.

His forecast was correct, for the comet did appear in that year, but Halley did not see it; he had died in 1742. For the first time scientists realized that comets can be regular visitors, and the comet was named after Halley. It takes the comet about 76 years to complete each trip around the Sun—which is fairly short for a comet. Last appearing in 1986, Halley's comet will not be seen again until 2061.

Structure and Composition

The only permanent part of a comet is its solid nucleus, or core. Other features of comets are the gaseous coma and the tail.

THE NUCLEUS

The nucleus of a comet is typically very dark, irregularly shaped, and several miles in diameter. It is often described as a "dirty snowball" because it consists of ice mixed with large amounts of fine, sooty dust particles. Some comets have more dust than ice. The ice in

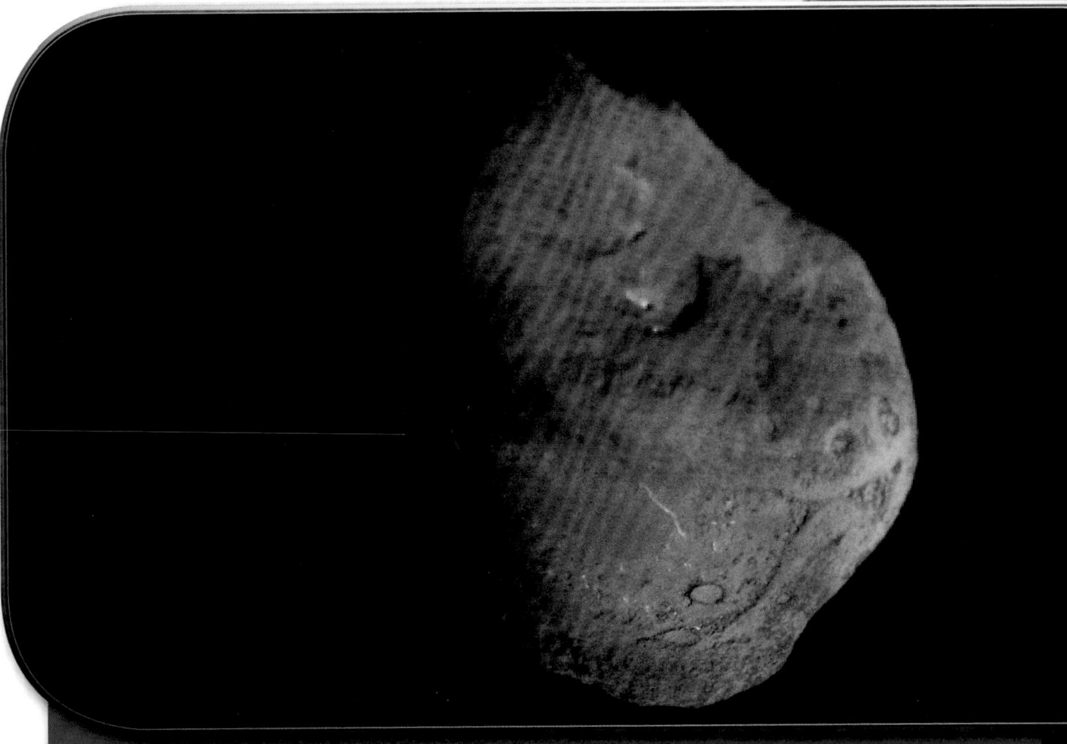

Comet Tempel 1's nucleus appears in a composite of several images taken by the Deep Impact spacecraft. The nucleus measures about 9 miles (14 kilometers) at its widest point. It has a powdery surface and a variety of terrain, including smooth and rough areas and what appear to be impact craters. NASA/JPL/UMD

a comet is mainly frozen water, with smaller amounts of frozen carbon monoxide, carbon dioxide, methane, ammonia, and other frozen gases. The dust contains rocky material and organic compounds. Comet nuclei are fragile and have been observed breaking up into fragments.

The Coma

As a comet nears the Sun, the ice in the nucleus begins to sublimate—that is, to pass directly from a solid to a gas. The gas carries with it some of the loosely bound dust particles. The gases spread out around the nucleus, forming a huge, dusty atmosphere called the coma. The nucleus and coma together make up the head of the comet. The diffuse, gaseous coma is what makes the head of a comet appear hazy. The coma is enormous, typically reaching about 60,000 miles (100,000 kilometers) or more in diameter. Sunlight causes the atoms in the coma to glow. If the supply of gases from the nucleus changes, a comet can brighten or fade unexpectedly, so astronomers cannot predict how bright a comet will become.

THE TAIL

As a comet approaches the Sun, radiation from the Sun usually blows dust from the comet into a dust tail. The tail is typically wide, slightly curved, and yellowish. The solar wind, a stream of highly energetic charged particles from the Sun, often sweeps hot gases away in a slightly different direction, producing another tail. It is usually fairly narrow, straight, and bluish. This tail is formed of plasma, or gases heated so much that they are electrically charged, with the electrons stripped away from the atomic nuclei. Comet tails may extend roughly 60 million miles (100 million kilometers) or more, but they contain only a small amount of matter. They point generally away from the Sun because of the force exerted by radiation and the solar wind on the cometary material. When comets travel away from the Sun, therefore, their tail or tails are in front of them.

DEAD COMETS

Each time a comet passes close to the Sun, it loses some of its matter. Eventually, the comet may disintegrate, ending up as only

The doomed Comet SOHO-6, at bottom left, streaks toward the Sun, where it ultimately burned up. In the image, taken by the Solar and Heliospheric Observatory (SOHO) spacecraft, the bright disk of the Sun has been blocked out to reveal the Sun's much fainter outer atmosphere. The white circle indicates the Sun's size and position. SOHO-6 was one of a group of "Sun-grazing" comets thought be fragments of a single, larger comet that broke apart. NASA/ESA/SOHO

a swarm of particles. Alternately, all the ices may eventually vaporize away from near its surface, leaving a dormant, or dead, comet, which resembles an asteroid. Perhaps 20 percent of near-Earth asteroids are thought to be dead comets.

ORBITS AND SOURCES

Comets orbit the Sun in elliptical, or oval-shaped, orbits that tend to be highly eccentric, or elongated. A comet's distance from the Sun usually varies considerably along its orbit. Based on their orbits, comets can be divided into two main types: short-period comets and long-period comets.

Short-period comets take less than 200 years to complete one orbit around the Sun. Most of them take less than 20 years and are called Jupiter-family comets. Long-period comets take between 200 and a million years to orbit the Sun. More distant comet nuclei—those that are more than 10,000 times farther from the Sun than Earth is—have probably never been inside the planetary system before. When one is dislodged from such a great distance, perhaps as far away as halfway to the nearest stars, it arrives in the inner solar system as a new

comet. The orbits of long-period and new comets are often extremely elongated and greatly inclined relative to the plane in which the planets orbit. Moreover, about half of them orbit the Sun in the direction opposite

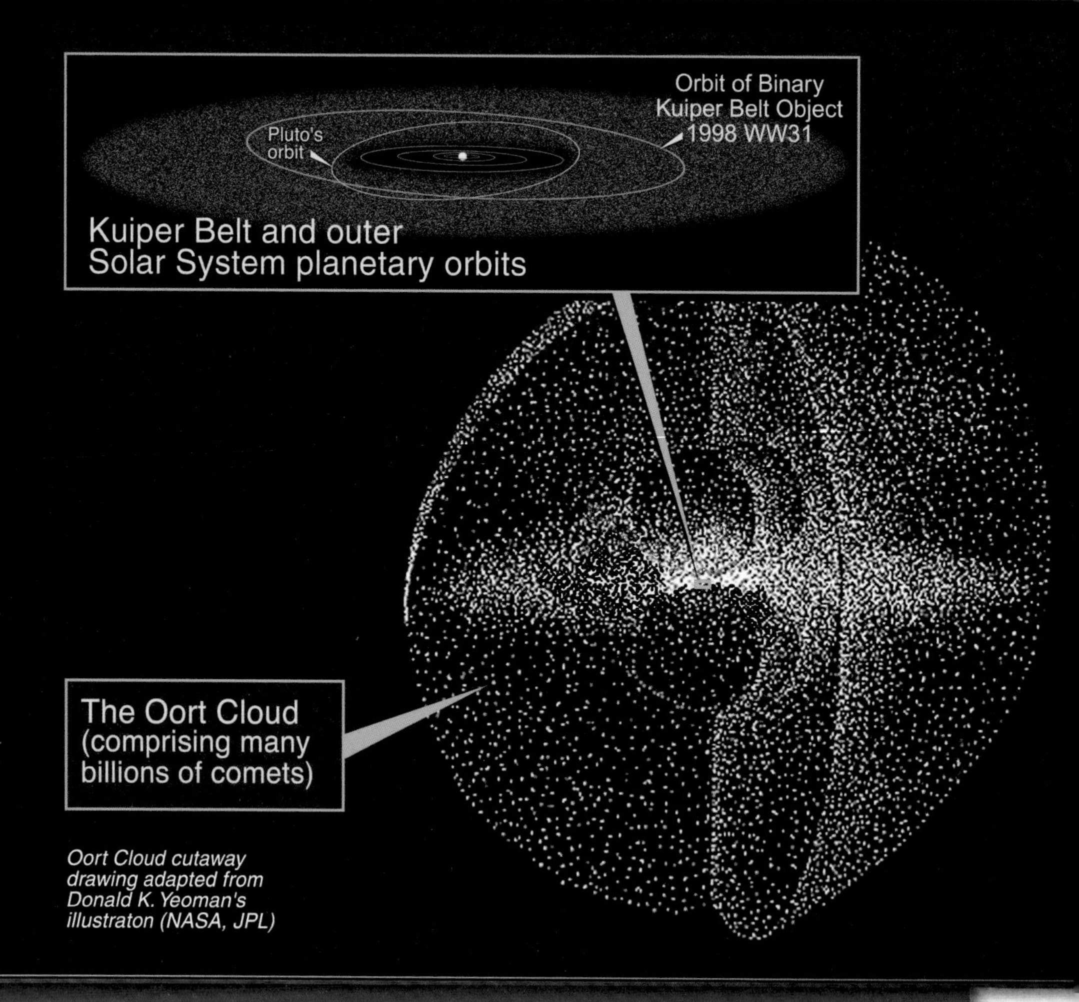

Oort Cloud. **NASA and A. Feild (Space Telescope Science Institute)**

that of the planets. Short-period comets usually have more circular orbits that lie within the plane of the planets' orbits. Most of them orbit in the same direction as the planets.

These two different types of comets apparently come from two different sources—the distant regions called the Kuiper belt and the Oort Cloud. Each region is a vast reservoir of comet nuclei, consisting of countless icy small bodies orbiting the Sun. Most short-period comets are thought to originate in the Kuiper belt. The more-distant Oort Cloud is the source for most long-period comets. It is a spherical cloud of comet nuclei orbiting the Sun in all directions. Sometimes the gravity of a larger body may alter the orbit of a comet nucleus in these regions, sending it on a path that takes it closer to the Sun. The object then becomes a comet.

Spacecraft Exploration

Many spacecraft missions to comets have been highly successful. In the 1980s several probes, including the European Space Agency's Giotto, flew by Halley's comet. Giotto was the first mission to return close-up images of a comet's nucleus. Deep Space 1, a probe sent by NASA, flew near

Comet Borrelly in 2001 and took detailed images of its nucleus. In 2005 NASA's Deep Impact craft intentionally crashed a projectile into the nucleus of Comet Tempel 1 to study the ejected debris, which provided

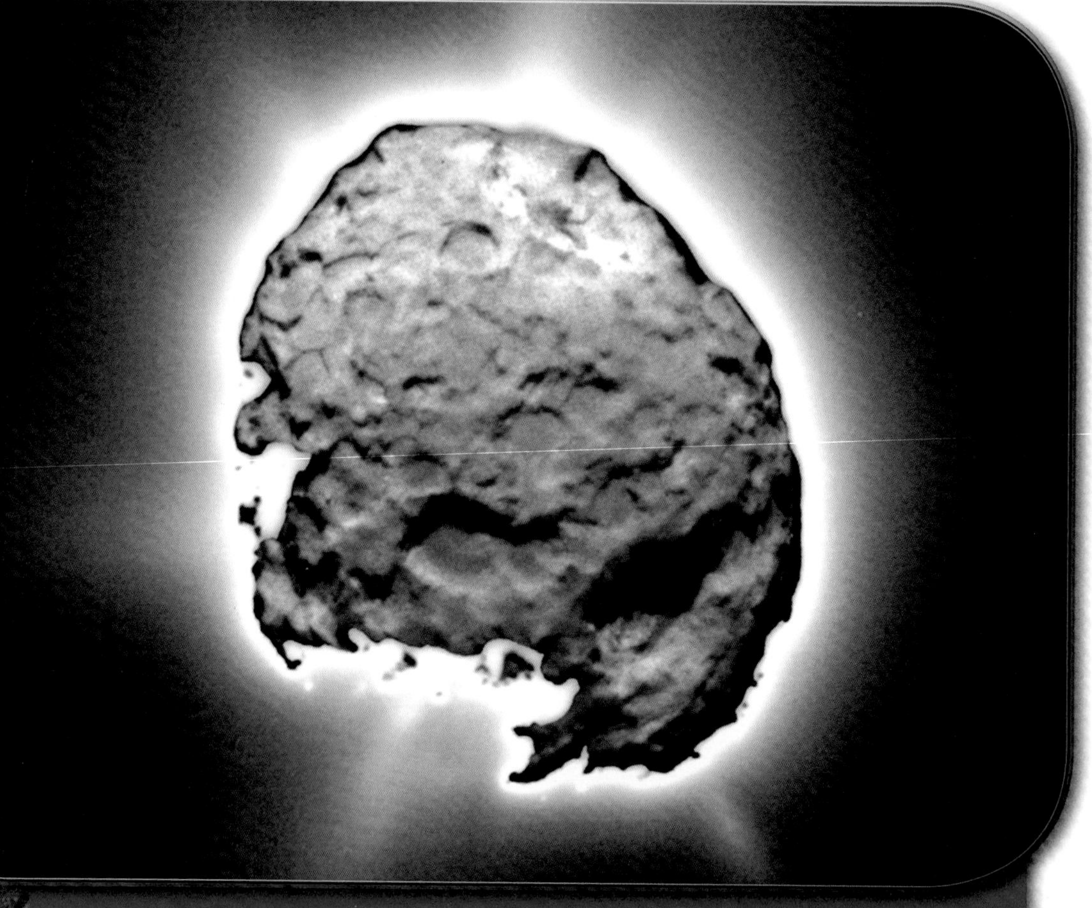

The Stardust spacecraft took this composite image of Comet Wild 2's nucleus during a flyby in 2004. It combines a short-exposure image that resolved surface detail and a long-exposure image that captured jets of gas and dust streaming away into space. NASA/JPL-Caltech

clues to the comet's composition. The mission captured images of the nucleus before, during, and after the impact. The surface of Tempel 1 contained a large amount of extremely fine, powdery dust that was only weakly held together by gravity.

NASA's Stardust probe visited Comet Wild 2 in 2004, photographing its nucleus and collecting samples of particles from its coma. The probe successfully returned the samples to Earth in 2006, allowing scientists to study a comet's material directly for the first time. Early analysis of the samples revealed a surprise: a grain of a mineral that formed under very high temperatures near the Sun during the early history of the solar system. Comets are believed to have formed in the colder,

Fragments of Comet Shoemaker-Levy 9 appear in a composite of images taken by the Hubble Space Telescope in 1994. A close encounter with Jupiter in 1992 broke up the comet's single nucleus into more than 20 pieces. The pieces lined up along the comet's orbital path, resembling a string of pearls. They eventually crashed into Jupiter's atmosphere. NASA/STScI/H.A. Weaver and T.E. Smith

outer part of the developing solar system. Before studying the Wild 2 sample, scientists thought that comets could not have come into contact with material from the inner part of the early system.

Another type of opportunity to study a comet came in the 1990s, when Comet Shoemaker-Levy 9 passed near Jupiter and broke into many fragments, which crashed spectacularly into the planet. The collisions temporarily left dark spots the size of Earth in Jupiter's atmosphere. NASA's Galileo spacecraft captured images of the event, as did Earth-based and Earth-orbiting telescopes.

6 chapter

Meteors and Meteorites

A flaming streak flashes across the night sky and disappears. On rare occasions the flash of light plunges toward Earth, producing a boom like the thundering of guns and causing a great explosion when it lands. When ancient peoples witnessed such displays, they believed they were seeing a star fall from the sky, and so they called the object a shooting star or a falling star.

Today these blazing trails of light are more fully understood. They are known to be caused by small chunks of stony or metallic matter from outer space that enter Earth's atmosphere and vaporize. Before they encounter Earth's atmosphere, these chunks of matter are called meteoroids. Once they enter the atmosphere, they are called meteors. Most meteors never reach Earth—they are so tiny that they vaporize completely soon after entering the atmosphere. Sometimes the particles are large enough, however, that they remain partly intact. The large, dense objects that survive the fall to Earth are called meteorites. Although thousands of meteoroids enter the atmosphere each year,

A Leonid meteor shower. **Jamal Nasrallah/AFP/Getty Images**

it is estimated that only about 500 actually reach the ground before vaporizing.

Classification of Meteorites

Although they come from outer space, meteorites consist of the same chemical elements as terrestrial matter. These elements, however, exist in meteorites in markedly different proportions from materials of Earth. They coalesce in characteristic ways to form the fabric of meteorites: either a metallic alloy of iron and nickel or a stone rich in silicon and oxygen.

There are three distinct groups of meteorites, classified according to their composition. Those in the first group are composed of the iron-nickel alloy and are called iron meteorites. The second type are made of stone and are called stony meteorites. Those in the third group are a mixture of stone and metal and are called stony-iron meteorites. A trained observer can identify iron meteorites by sight, by chemical testing, or by etching with acid, which causes a characteristic pattern of crisscross bands. Stony meteorites are the ones most commonly observed falling to Earth. They are far more difficult to identify by sight, however, because to the untrained eye they resemble other terrestrial stones. Nevertheless, chemical tests and X-ray examinations can positively identify a true stony meteorite.

A Meteor's Journey to Earth

A meteoroid moving through outer space may pass close enough to Earth to be trapped

by the planet's gravitational field. If this happens, the meteoroid is drawn into the atmosphere and toward Earth's surface by the force of gravity.

When a meteoroid enters the atmosphere, it is traveling at a tremendous speed—from 1,100 to 5,200 miles (1,800 to 8,400 kilometers) per hour—much faster than the surrounding air. This not only causes a sonic boom from the resulting shock wave but also produces frictional heat high enough to raise the meteoroid's surface to its boiling point. The surface material then simply melts away, or vaporizes, resulting in a fine dust. Most meteors are very small and are vaporized completely in this way. Millions of these burned-out particles fall to Earth as dust every day.

The outer surfaces of even large meteors melt rapidly. As passing airstreams sweep away the melted upper layer, a new layer is exposed and melted in turn. The melting surface gives off glowing particles that stream behind the meteor as it speeds through the atmosphere, creating a blazing trail.

The meteor's flight through Earth's atmosphere lasts only a few seconds. If the body is large, the heat at its surface does not have time to penetrate deeply into the interior.

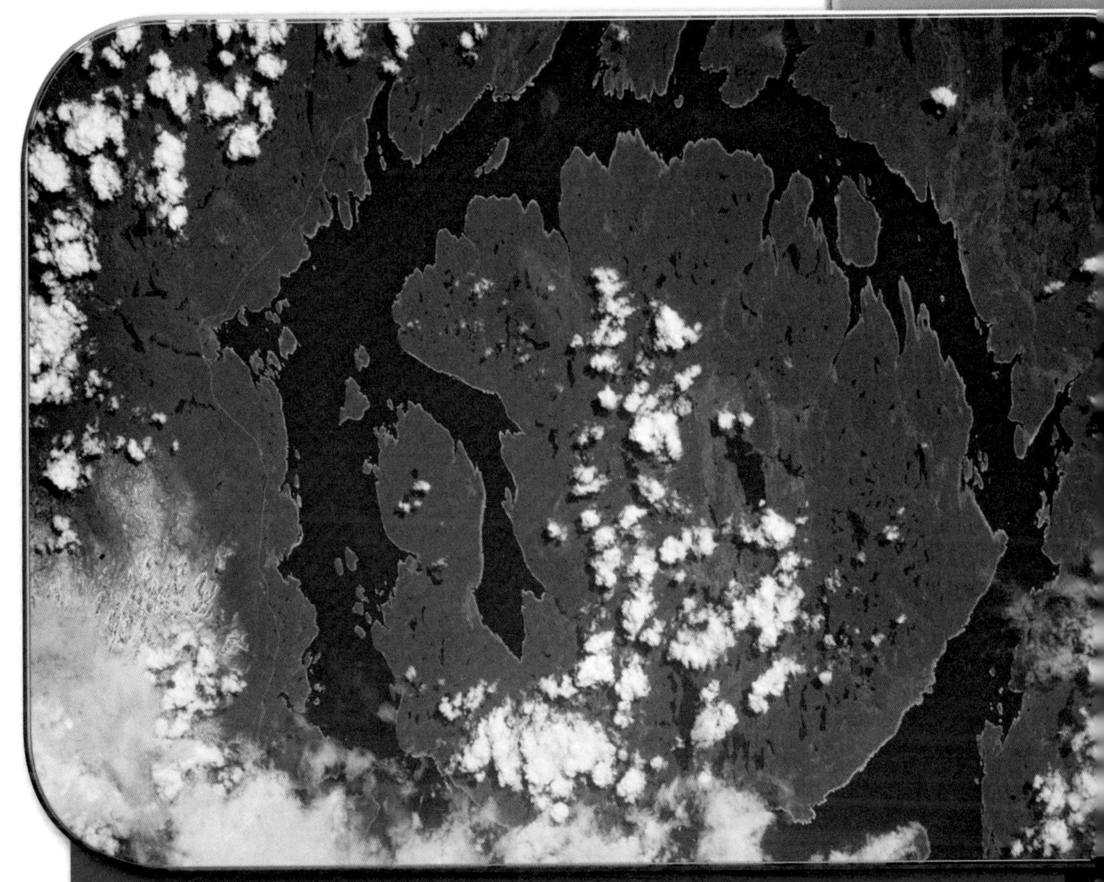

The Manicouagan Reservoir Crater in Quebec, Canada, is one of the largest meteorite craters in the world. It was formed when Earth was blasted by a giant meteorite at the end of the Triassic period some 210 million years ago. A mass extinction of marine species occurred about the same time, and scientists believe that sky-darkening dust from the meteorite's impact may have played a role. NASA

As the meteor approaches Earth's surface, air resistance slows it. In the dense lower atmosphere, the body often bursts into fragments to produce a meteor shower. Rarely,

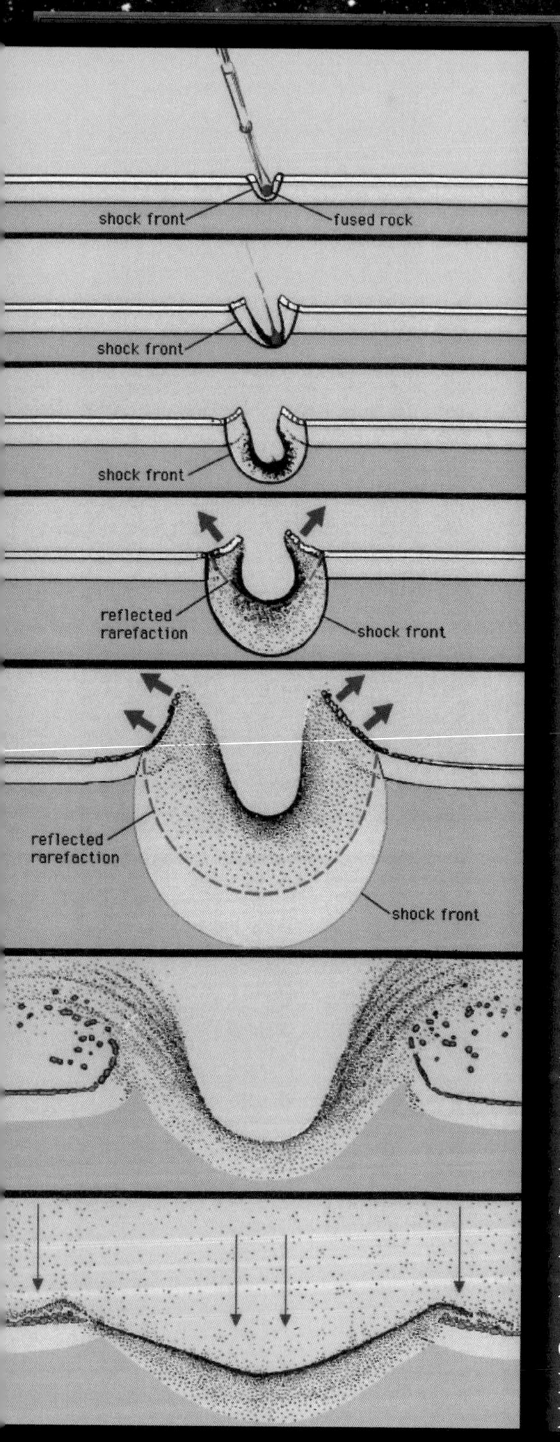

Meteorite Craters

The series of illustrations shows the sequence of events thought to occur in the formation of a simple meteorite crater. The impact of the meteorite striking the ground creates shock waves, which move at a higher velocity than the existing elastic waves in the ground. A spherical shock front in the rocks outruns the penetrating meteorite and forms an expanding cavity. As the pressure behind the shock front rises, the rocks behind are strongly compressed. Depending on the properties of the rocks, segments are vaporized, melted, crushed, fragmented, or fractured. They are ultimately pushed outward and ejected, generally in a radial pattern. The increased density of the compressed rocks behind the shock front lasts until rarefaction, or decompression, waves allow progressive relaxation.

There are many large craters on Earth's surface that are known to be the result of meteorite impacts. They have been found in all parts of the world—from central Australia to the Arabian Desert.

very intense storms, such as the great Leonid showers of 1833 and 1966, rain tens of thousands of meteors over a short period of time. If a meteor remains intact, however, the fireball dies and the melted surface solidifies into a dark crust before the meteor has a chance to hit the ground.

Most meteorites break into small particles when they strike Earth, but on rare occasions they do not. One example, the largest known meteorite, is the Hoba West, which weighs about 60 tons. It was discovered many centuries after it had fallen to Earth in Namibia. The second largest meteorite is the Ahnighito, which weighs about 34 tons. It was discovered in Greenland.

The Origin of Meteors

As Earth travels in its orbit around the Sun, it continually encounters meteoroids head-on. On a clear, dark night an observer may see 10 or more meteors per hour. Sometimes an unusually large number of small meteors can be seen in rapid succession—perhaps more than 50 per hour. Such a display is called a meteor shower and occurs when Earth passes through a swarm of meteoroids. Because of their small size, these meteors generally burn

Barringer (or Meteor) Crater, in the Canyon Diablo region of Arizona, was discovered in 1891. The crater is 600 feet (180 meters) deep and 4,000 feet (1,200 meters) wide. Scientists estimate that a small asteroid about 150 feet (45 meters) in diameter created the hole some 25,000 years ago. **D.J. Roddy/U.S. Geological Survey**

up in the upper atmosphere and never reach the ground.

Some meteor showers occur regularly each year and coincide with the passage of Earth through the orbit of a comet. As a comet moves in its orbit, it leaves a trail of debris.

This debris may be tiny pieces of grit and ice that have escaped from the tail of the comet, or it may be the fragmentary remains of a comet that has disintegrated. Some astronomers suggest that the tiny meteoroids that cause meteor showers may actually be this cometary debris.

However, meteorites—the meteors that do reach the ground—have a composition similar to asteroids, and their orbits resemble asteroidal orbits. Scientists have determined that over 99 percent of meteorites are fragments of asteroids. Less than 1 percent are thought to come from the Moon or Mars. There is also reason to believe that some meteorites are fragments of the rocky remnants of comets, though this remains to be firmly established.

Conclusion

The skies have long fascinated scientists and laypeople alike. Ancient peoples made the first discoveries of planets with the unaided eye. Later, telescopes allowed astronomers to locate all the planets of the solar system as well as hundreds of thousands of other objects orbiting the Sun. Yet, the very nature of discovery means that there is always more to be learned. In other words, the process of discovery never ends. Even as improvements in technology enable astronomers to learn more about the solar system, new questions are raised for future research.

Sometimes this research can force astronomers to revise long-accepted knowledge. The reclassification of Pluto as a dwarf planet in 2006 is a recent example of how new discoveries can change what we know about space. The ongoing discoveries of extrasolar planets, as well as missions to distant asteroids and comets, promise to continue to expand our knowledge of the skies.

accrete To grow through the gradual accumulation of material.

almanac A publication containing astronomical and meteorological data for a given year and often including miscellaneous other information.

ammonia A pungent (strong-smelling), colorless, gaseous alkaline compound of nitrogen and hydrogen that is quite soluble in water and easily condensed to a liquid by cold and pressure.

carbonaceous Relating to, containing, or composed of carbon.

cavity An unfilled space within a mass; a hollowed-out space.

coalesce To unite into a whole; fuse.

decompression The release from pressure or compression.

deviation An act or instance of straying, especially from a standard, a principle, or an established course.

discord Lack of agreement or harmony.

discrepancy The quality or state of disagreeing or being at variance.

dissipate To spread thin or scatter and gradually vanish.

dormant Inactive.

ecliptic The plane of Earth's orbit around the Sun.

elongated Stretched out.
moonlet A small natural or artificial satellite.
obscure To make dark, dim, or indistinct; to conceal or hide.
occultation The interruption of the light from a celestial body caused by another celestial body passing in front of it.
planetesimal Small, solid heavenly bodies that are theorized to have coalesced to form Earth and the other planets early in the history of the solar system.
primordial Existing in or persisting from the beginning.
protoplanet A hypothetical eddy within a giant, whirling cloud of gas and dust that becomes a planet by condensation during formation of a solar system.
rigor Strict precision; exactness.
serendipitous Obtained or characterized by serendipity, or a lucky accident.
silicaceous Relating to, containing, or composed of silica.
subtle Delicate, elusive; difficult to understand or perceive.
synchronous Happening, existing, or arising at precisely the same time.
tenuous Having little substance or strength.
vicinity A surrounding area.

American Astronomical Society
2000 Florida Avenue NW
Suite 400
Washington, DC 20009-1231
(202) 328-2010
Web site: http://aas.org
The American Astronomical Society is a major organization of professional astronomers in North America. Its members also include physicists, mathematicians, geologists, engineers, and others interested in contemporary astronomy.

Canadian Space Agency
John H. Chapman Space Centre
6767 Route de l'Aéroport
Saint-Hubert, QB J3Y 8Y9
Canada
(450) 926-4800
Web site: http://www.asc-csa.gc.ca/
The Canadian Space Agency provides resources for students, teachers, and the scientific community about space and space exploration. The Web site offers biographies of Canada's astronauts.

National Aeronautics and Space Administration (NASA)
The Space Place

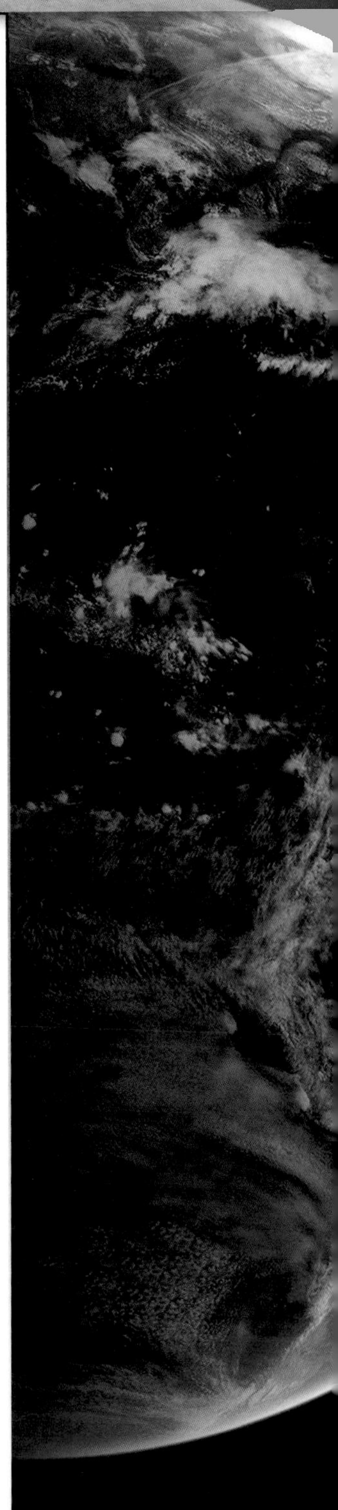

New Millennium Program Education and Public Outreach
Jet Propulsion Laboratory
Mail Stop 606-100
4800 Oak Grove Drive
Pasadena, California 91109
Web site: http://spaceplace.jpl.nasa.gov/en/kids/index.shtml
NASA's Space Place is a joint effort by the National Aeronautics and Space Administration, Jet Propulsion Laboratory, the California Institute of Technology, and the International Technology and Engineering Education Association. The site provides space- and technology-related games, projects, animations, and facts for students.

Royal Astronomical Society of Canada
203 - 4920 Dundas Street W
Toronto, ON M9A 1B7
Canada
(416) 924-7973
Web site: http://www.rasc.ca
The Royal Astronomical Society of Canada is the country's foremost astronomy organization. It unites amateurs, educators, and professionals and offers local programs and services throughout Canada.

Windows to the Universe
National Earth Science Teachers Association
PO Box 3000
Boulder, CO 80307
(720) 328-5350
Web site: http://windows2universe.org/our_solar_system/solar_system.html
This Web site for teachers and students provides articles on numerous elements of the solar system, including the planets, dwarf planets, asteroids, meteors, and comets.

Web Sites

Due to the changing nature of Internet links, Rosen Educational Services has developed an online list of Web sites related to the subject of this book. This site is updated regularly. Please use this link to access the list:

http://www.rosenlinks.com/tss/pdpso

Bibliography

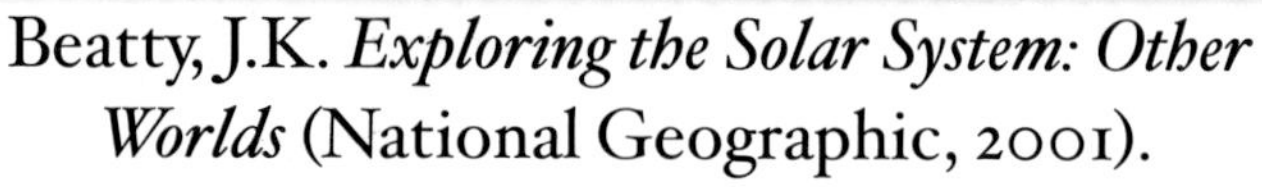

Beatty, J.K. *Exploring the Solar System: Other Worlds* (National Geographic, 2001).

Bell, Jim, and Mitton, Jacqueline, eds. *Asteroid Rendezvous: NEAR Shoemaker's Adventures at Eros* (Cambridge Univ. Press, 2002).

Croswell, Ken. *Ten Worlds: Everything That Orbits the Sun* (Boyds Mills, 2007).

Elkins-Tanton, L.T. *Asteroids, Meteorites, and Comets* (Chelsea House, 2006).

Hartmann, W.K., and Miller, Ron. *The Grand Tour: A Traveler's Guide to the Solar System*, 3rd ed. (Workman, 2005).

Koppes, Steven N. *Killer Rocks from Outer Space: Asteroids, Comets, and Meteorites* (Lerner, 2004).

Miller, Ron. *Asteroids, Comets, and Meteors* (Twenty-First Century Books, 2006).

Ride, Sally, and O'Shaughnessy, T.E. *Exploring Our Solar System* (Crown, 2003).

Sumners, Carolyn, and Allen, Carlton. *Cosmic Pinball: The Science of Comets, Meteors, and Asteroids* (McGraw Hill, 2000).

Villard, Ray, and Cook, Lynette. *Infinite Worlds: An Illustrated Voyage to Planets Beyond the Sun* (Univ. of Calif. Press, 2005).

Weintraub, D.A. *Is Pluto a Planet?: A Historical Journey Through the Solar System* (Princeton Univ. Press, 2009).

A

apparent motions, 36–37
asteroids, 14, 58–65, 73, 87
  near-Earth and potentially hazardous, 62–63
  physical characteristics of, 58–62
  spacecraft exploration of, 63–65

C

Ceres, 14, 39, 41, 58, 59, 65
Charon, 42, 45, 51–53, 54
collisions, 20–22
comets, 66–78
  dead, 71–73
  orbits and sources, 31, 73–75, 86–87
  spacecraft exploration of, 75–78
  structure and composition of, 68–73

D

dwarf planets, 14, 39–57

E

Earth
  meteors/meteorites and, 79–80, 81–86
  orbit and rotation of, 29, 34, 35–37, 46
Eris, 13, 14, 31, 39, 41, 54–57
extrasolar planets, 25–28

H

Halley's comet, 66, 68, 75
Hubble Space Telescope, 27, 45, 54

I

inner planets, 14–16, 17, 19–20
International Astronomical Union (IAU), 13–14, 39, 41

J

Jupiter, 10, 11, 17, 58, 60, 78
  rotation of, 35

K

Kepler, Johannes, 29–31, 32–33
Kuiper belt, 13, 14, 31, 39, 44, 45, 54, 56, 75

L

laws of planetary motion, Kepler's, 29–31, 32–33

M

Mars, 10, 11, 15, 16, 58, 87
  orbit and rotation of, 32, 34, 36
Mercury, 10, 11, 15, 21–22, 23, 25, 40–41, 45, 46
  orbit and rotation of, 29, 31, 34–35, 37, 46
meteors and meteorites, 63, 64, 79–87
  classification of, 80–81
  craters of, 84
  journey to Earth, 81–85
  origin of, 85–87
Moon, 11–12, 16, 20–21, 23, 38, 45, 50, 51, 54, 87

N

NASA, 45, 63, 65, 75–77
Neptune, 10, 12, 13, 17, 41, 45, 66
  orbit and rotation of, 31, 36, 43, 46, 54

O

outer planets, 15, 17, 19–20, 24

P

planets, 10–14
  characteristics of, 13–14
  classification of, 14–17
  formation and evolution of, 17–25
  formation of layers, 22–23
  loss of internal heat, 23–25
  motions of, 29–38
  outside the solar system, 25–28
Pluto, 42–53, 55
  atmosphere, surface, and interior of, 48–51
  defining as planet/dwarf planet, 10, 12–13, 14, 39, 41, 44
  moons of, 42, 45, 51–53, 54
  orbit and rotation of, 29, 31, 46–48, 54
  study of, 44–45
plutoids, 41, 54

S

Saturn, 10, 11, 17
  orbit and rotation of, 36, 37

U

Uranus, 10, 12, 17, 22, 66
  rotation of, 33, 36, 42–43, 47

V

Venus, 10, 11, 15, 22, 24
  orbit and rotation of, 31, 33, 34–35, 37